Novel Approaches in Materials and Technologies for Building Construction

Mohammad Arif Kamal
Rupesh Manohar Surwade
Abhishek Mahendra Bangre

Published by **Materials Research Forum LLC**
Millersville, PA 17551, USA

Published as part of the book series
Materials Research Foundations
Volume 158 (2024)
ISSN 2471-8890 (Print)
ISSN 2471-8904 (Online)

Print ISBN 978-1-64490-292-9
ePDF ISBN 978-1-64490-293-6

This book contains information obtained from authentic and highly regarded sources. Reasonable efforts have been made to publish reliable data and information, but the authors and publisher cannot assume responsibility for the validity of all materials or the consequences of their use. The authors and publishers have attempted to trace the copyright holders of all material reproduced in this publication and apologize to copyright holders if permission to publish in this form has not been obtained. If any copyright material has not been acknowledged, please write and let us know so we may rectify in any future reprint.

Distributed worldwide by

Materials Research Forum LLC
105 Springdale Lane
Millersville, PA 17551
USA
https://www.mrforum.com

Printed in the United States of America
10 9 8 7 6 5 4 3 2 1

Table of Contents

Preface

Technological progress has introduced many innovations and technologies in the field of building industry. Building construction technology is concerned with the technical performance of buildings, building materials, and building construction systems. The building technology encompasses; the functioning of components and systems; the principles; procedures, methods and details of building assembly; operating strategies; materials and their applications, management etc. Built on a legacy of thousands of years of gradual innovation in construction technology and scientific advancement, architectural engineering applies the latest scientific knowledge and technologies to the design of buildings. The building construction industry is continuously being updated and upgraded with the latest technologies to improve the cost, safety, efficiency, and quality of construction. The use of machinery and automation has made its way through the civil engineering and construction industry. Most of the building components such as columns, roofs and concrete blocks are available as prefabricated forms that increase the speed of construction process greatly. From reduction of greenhouse gas emissions to the construction of resilient buildings, architects and building engineers are at the forefront of addressing several major challenges of the 21st century. Building Technology adds significant value at every stage of construction, from planning to commission, by optimizing efficiency and productivity. The goal is to engineer high-performance buildings that are sustainable, efficient, resilient, and economically viable, that ensure the safety, health, comfort, and productivity of occupants. In the rapidly changing scenario of building sector, architects, engineers and builders should search for new materials and building construction technologies to adopt in future constructions that benefits like energy efficiency, resources and water conservation, improved indoor air quality, life cycle cost reduction, durability and low maintenance.

This book 'Novel Approaches in Materials and Technologies for Building Construction' contains eleven chapters that introduces and discusses some state-of-the-art building technologies and scientific methods that may be beneficial to building engineers, architects, students of civil engineering and architecture, research scholars, building scientists and construction industry professionals.

CHAPTER 1

Calcium Silicate Bricks

1.   Introduction

Calcium silicate bricksare made of sand and lime and are popularly known as sand-lime bricks. These bricks are used for several purposes in construction industries such as ornamental works in buildings, masonry works, etc. Sand lime bricks are popularly used in European countries, Australia, and African countries. In India, these bricks are widely used in Kerala state and their usage is regularly growing. Calcium silicate bricks are suitable for use in both external and protected internal walling. They are available as facing bricks or as commons. As for clay bricks, the bricks are available in a solid or a frogged unit and are made to a standard size of 216 x 102.5 x 65 mm. The method of manufacture together with inherent properties of the mixed raw materials produces a brick with fine dimensional tolerances and good clean arises. Figure 1 shows the colored Calcium Silicate or Sand Lime Bricks [1].

Figure 1. The colored Calcium Silicate or Sand Lime bricks

2.   Raw Materials Used for Sand Lime Bricks

The raw materials used in the manufacture are a very fine siliceous aggregate, high calcium lime, and water. Inert and stable pigments are normally added to give the required color. The materials are first mixed in the required proportions and are mechanically pressed under considerable pressure into molds. They are then cured in high-pressure steam autoclaves for several hours which results in the combination of the lime with part of the siliceous aggregate to produce a hydrous calcium silicate (tobermorite) which forms the binding medium in the finished brick. The materials listed below are used for the production of calcium silicate bricks [1].

2.1　Sand

Calcium silicate bricks contain a high amount of sand is about 88 – 92%. It means the properties of these bricks depend upon the characteristics of sand used. Sand has two distinct functions to perform in the manufacture of sand-lime brick, which require different properties of the material. Part of it must enter into a chemical combination with the lime to form the calcium-silicate bonding material. The rest of the sand grains constitute the aggregate which is bound together and which forms the main body of the brick. It is necessary, therefore, that that part of the sand which is expected to combine with the lime shall be in as fine a state of division as possible. If this does not occur naturally, some of the sand must be ground until it is of sufficient fineness.A good practical rule is that about 15% of the sand must pass a 100-mesh screen. The remaining 85 % of the sand is intended to form the inert filler or main body of the brick.The sand used shall be well-graded and should not contain any impurities like organic matter, soluble salts, etc. the finely divided clay may be present but it is up to 4% only which helps the brick in pressing and provides smoother texture.

2.2　Lime

Lime content in calcium silicate bricks varies from 8 to 12%. The lime used shall be of good quality and high calcium lime. Although the proportion of lime used in making sand-lime brick is relatively small, its quality is of paramount importance. The lime must be perfectly hydrated before the bricks are pressed. Otherwise, it will expand during the steam treatment and produce internal strains which are frequently sufficient to disrupt the brick. The lime must also be sufficiently caustic to enter readily into combination with the sand

2.3　Water

Clean water should be used for preparing calcium silicate bricks. Seawater or water containing soluble salts or organic matter of more than 0.25% is not suitable.

2.4　Pigment

Pigments are generally used to give color to the bricks. They are added to the sand and lime while mixing. The total weight of brick contains 0.2 to 3 % of pigment quantity. Different pigments used to get different colors are tabulated in Table 1.

Table 1. The different constituent materials used for Sand Lime bricks

Pigment	Color
Carbon black	Black, grey
Iron oxide	Red, brown
Chromium oxide	Green
Ochre	Yellow

3.   History of Calcium Silicate Bricks

The process for making brick of sand was discovered and patented by Dr. William Michaelis in 1880. Peppel indeed recognizes sand-lime brick made by other processes, and numerous statements may be found in the literature concerning a "sand-lime brick" made in New Jersey about i860. This material was a mortared brick, being made of ordinary lime mortar, molded into shape, and permitted to set. So far as can be learned, none of these earlier processes has proved a commercial success. All sand-lime brick now marketed in this country is made according to the basic principles covered by the original Michaelis patent. Dr. Michaelis permitted his patent to lapse without exploitation, almost immediately thereafter several modifications.

4.   Manufacturing of Calcium Silicate Bricks

Calcium silicate (sand lime and flint lime) bricks are manufactured by mixing lime, sand, and/or crushed silicaceous or flint stone together, with enough water to allow the mixture to be molded under high pressure. The bricks are then steam autoclaved so that the lime reacts with the silica to form hydrated calcium silicates. Pigments can be added during the mixing stage. In their natural state, calcium silicate bricks are white to a creamy off-white color, but the addition of ochres (buff or cream colors), iron oxides (pink, red, brown, or black), or chrome oxide (green) can enable a very wide variety of colors to be produced [2].

4.1   Proportioning of Materials

The relative proportions of sand and lime which are used in the manufacture of sand-lime brick are of course very important in determining the properties of the finished product. In the first step, suitable proportions of sand, lime, and pigment are taken and mixed thoroughly with 3 to 5 % of water. Then paste with moldable density obtains. The mixture is molded into bricks using a rotary table press which uses mechanical pressure to press the bricks. The pressure of pressing varies from 31.5 to 63 N/mm^2.

It will be remembered that Michaelis's patent called for "from 10 to 40 parts of calcium hydroxide to 100 parts of sand."The pure calcium silicate is a gelatinous material which shrinks on drying like lime paste. To obtain a lime mortar of maximum strength, it is necessary to add enough sand to the paste so that the internal strains caused by this shrinking will be largely overcome.The hydrate is therefore 25 percent by volume of the sand. Assuming that sand weighs 100 pounds per cubic foot and hydrated lime weighs 40, the hydrate is about 10 percent by weight of the sand. If the hydrate carries 24 percent water the above proportions are equivalent to 7.6 parts of quicklime to 100 parts of sand by weight.

4.2   Mixing of Materials

Probably the most important step in the manufacture of sand-lime brick is the mixing of lime and sand. This operation is usually the determining factor in the quality of the brick. Every precaution should be taken that the mixing is thorough and efficient so that the best results be obtained from the raw materials at hand. The lime and fine sand should be in intimate contact with each other so that the chemical reaction between them can readily take place. The coarse sand should be evenly distributed throughout the mass so that the proportion of voids shall be a minimum.

4.3   Compaction of Materials

The mixture of lime and sand is now ready to be pressed into the form of a brick. Pressing serves not only to give the brick its final size and shape but performs several more important functions. By bringing the sand and lime into very intimate contact with each other, the chemical combination between them can be facilitated. The compression of the material necessarily decreases the proportion of voids and therefore produces a less porous brick. The final strength of the brick has been found to depend, to some extent, on the pressure exerted in molding it. A sand-lime brick is not subject to any other change of shape or size after it leaves the press.

4.4   Hardening of Bricks

The bricks are picked by hand from the press table and piled upon iron trucks. This operation requires considerable care, for at this stage of the process the bricks are very tender and easily crushed in the hand. The trucks are designed to carry about 1100 bricks each. As soon as a truck is loaded it is pushed by hand on tracks into the hardening cylinder.The bricks are subjected to the steam treatment may be of almost any size or shape. In the final stage, bricks are placed in an autoclave. An autoclave is nothing but a steel cylinder with tightly sealed ends. The diameter of the autoclave is about 2m and the length is about 20m.This is a cylindrical shell of open-hearth steel, 70 feet long by 6 feet in diameter, built up of plates five-eighths of an inch thick, riveted together. After placing bricks in this closed chamber saturated steam pressure is released at about 0.85 to 1.6 N/mm^2.The temperature inside the chamber is raised and the reaction process begins.The shell is set up horizontally and tracks for the cars laid on the bottom of it. The front end is used as a cover and is held in position by 50 nickel-steel bolts 1inch in diameter. After erection, the cylinder is tested to a pressure of 225 pounds per square inch. Thirty-three such cylinders will hold 20 trucks, carrying 22,000 bricks. The cover of this cylinder is designed to be lifted off using a small chain block.The press is operated during the day until the cylinder is filled, and the steam treatment takes place during the night. Since it requires about 3 hours to bring the cylinder up to maximum pressure and about 1 hour to blow off the steam before the cover can be removed, the duration of curing is limited to about 10 hours. Under these circumstances, a steam pressure of about 120 pounds per square inch has been found satisfactory. Silica content in sand and calcium content in lime reacts and forms a crystal-like compound called calcium hypo silicate. This process is done for 6 to 12 hours. Finally, the obtained bricks are transported to the workplace.

Figure 2. The 20m long steel cylinder act as an Autoclave

4.5   Stacking and Handling of Bricks

When the trucks are taken out of the hardening cylinder, they should immediately be taken to the loading platform, where the bricks are transferred to freight cars for shipment. The bricks are generally handled by one of three methods:

a) Passing them by hand from one man to another;

b) The hand carrier, which is built on the principle of a pair of tongs and is capable of picking up 8 or 10 bricks

c) The gravity carrier is a chute set on small iron legs arranged to give it a slight inclination. The bed of the chute is made up of small wooden rollers, set close together, which are free to turn as the bricks slide over them.

Figure 3. The stacking of Calcium Silicate Bricks

5. Properties of Sand Lime Bricks

The following properties of sand-lime bricks are summarized as follows [3]:

a) They have a very smooth and uniform finish, it has an attractive appearance.

b) They are dense, strong, and tough.

c) They are porous therefore free from indigestion.

d) Also, uniform in size, shape, finish, and no plastering is required.

e) When required, the amount of plaster is significantly less.

f) Essential materials are quite common in the event can be used as an alternative to clay bricks

6. Advantages of Calcium Silicate Bricks

There are many advantages of calcium silicate bricks when used in masonry construction, and they are:

a) Mortar required for providing plaster on calcium silicate bricks is very less.

b) The color and texture of these bricks are uniforms.

c) The compressive strength of sand-lime bricks is about $10N/mm^2$. So, they are well suitable for multi-storied buildings.

d) For constructions in clay soils, these bricks are more preferable.

e) The trouble of Efflorescence does not arise in the case of sand-lime bricks.

f) Not only bricks, blocks, and tiles can also be made using calcium silicate.

g) Sand lime bricks provide more comfort and accessibility for architects to attain desired shape and designs.

h) These bricks have accurate shape and size with straight edges.

i) Solar heat effect is reduced on exposed walls made of calcium silicate bricks.

j) Colored sand-lime bricks do not need any finish to the wall, so, the cost reduces.

k) These bricks have great fire resistance and water repellant properties.

l) Calcium silicate brick walls resist noise from outside.

m) The cost of construction gets reduced by about 40% of the total cost due to the following factors.

n) Wastage of calcium silicate products is very less.

o) Less quantity of mortar is needed.

p) The thickness of the wall can be reduced when constructed using these bricks because of high compressive strength.

7.   Disadvantages of Calcium Silicate Bricks

In some conditions, calcium silicate bricks are not suitable and their disadvantages are:

a) If the clay is available in plenty, clay bricks are more economical than calcium silicate bricks.

b) These are not suitable for laying the foundation, because they cannot provide resistance to water for a longer period.

c) They cannot also resist fire for longer periods so, they are not suitable for building furnaces, etc.

d) The abrasion resistance of these bricks is very less so, they cannot be used as paving materials.

8.   Problems Associated with Calcium Silicate Bricks

Some problems are associated with Calcium Silicate bricks [4]. They are summarized as below:

a) Thermal movement is likely to be about 1.5 times that of clay brickwork. Calcium silicate brickwork, unlike clay, usually undergoes an initial irreversible shrinkage on laying (clay brickwork tends to expand) but so long as the propensity for movement is understood and catered for in the design, there is no reason why the brickwork should not perform adequately. Often this factor is not catered for in the design and this results in widespread cracking.

b) Calcium silicate bricks should not be used in solid work with clay facings or backings, this is because of the propensity of the bricks to shrink in contrast with the expansion of clay brickwork. If solid walling is to be contemplated, backings of concrete bricks or blocks should be used, as these have similar movement characteristics to calcium silicate bricks. We often see an inappropriate choice of walling material for the inner leaf and this sets up opposing forces due to differential expansion, again resulting in widespread cracking.

c) General construction detailing is often not attended to, particularly about providing sufficient flexibility in the wall ties to permit the differential movements and allowing for discontinuity around cavity closers to prevent cracking.

d) The requirement for inbuilt slip planes is often not attended to. Internally, walls of calcium silicate brickwork need to be bedded on a damp proof course to act as a slip plane and so facilitate longitudinal movements to occur – this would be equally necessary at upper floor levels, a detail that had been missed at this scheme.

e) Movement control in walling is not the only issue – also consider building elements that could provide a restraining influence. For example, concrete columns or walls cast up against bricks should be avoided unless a slip membrane can be provided. – as should any form of construction that will prevent free movement. At this scheme pointing of the movement joints and DPC, both provide this restraining influence.

f) It is not unusual to see some form of displacement with calcium silicate bricks due to thermal expansion, for example, brickwork sliding off a damp proof course, cracking at corners, or evident disruption. By contrast, shrinkage cracking does not generally produce these manifestations.

Conclusions

Calcium silicate bricks are often given bad press due to the issues highlighted here; however, it should be said that they are an excellent building material so long as the construction detailing required dealing with shrinkage or expansion is understood. Unfortunately more often than not this detailing is not understood and buildings are generally constructed in the same fashion as would clay bricks. They outperform clay bricks in some respects, particularly about frost resistance.

References

[1]   Information on https://theconstructor.org/building/calcium-silicate-bricks-masonry-construction/17256/

[2]   M. K. Sahu, L. Singh, Critical Review on Types of Bricks: Calcium Silicate Bricks, International Journal of Mechanical and Production Engineering, Vol. 5, Issue 11, 2017

[3]   Information on https://constructionor.com/sand-lime-bricks/

[4]   Information on http://buildingdefectanalysis.co.uk/masonry-defects/an-introduction-to-calcium-silicate-bricks/

Novel Approaches in Materials and Technologies for Building Construction   Materials Research Forum LLC
Materials Research Foundations **158** (2024)   https://doi.org/10.21741/9781644902936

CHAPTER 2

Autoclaved Aerated Concrete (AAC) Blocks

1. Introduction

Traditional bricks are the main building materials that are used extensively in the construction and building industry in India. Due to the rapid urbanization and expanding interest for development materials, block furnaces have quickly developed which have legitimately or in a roundabout way caused a progression of ecological and medical issues. At a worldwide level, ecological contamination from block-making activities adds to the wonders of an Earth-wide temperature boost and environmental change. The different kinds of squares can be utilized as an option in contrast to the red blocks, to diminish natural contamination and Global warming. AAC squares might be one of the answers for block substitution. Like froth concrete, Autoclaved Aerated Concrete (AAC) is one of the confirmed green structure materials, which can be utilized for business, modern and private development. The AAC block is a lightweight material and has high warm protection esteem. It has the basic properties required for use as a structured segment. Because of the lightweight and high solidarity-to-weight proportion of circulated air through solid items, their utilization brings about an obvious economy in the auxiliary individuals, and along these lines spare concrete and steel support. With AAC, the development procedure can be around 20 percent quicker. It weighs just around 50 percent of a standard solid square and has high warm protection and is acoustics-accommodating. It additionally has preferred imperviousness to fire over fly debris and is non-ignitable. It's non-unfavorably susceptible and consequently keeps up the nature of air inside a structure without changing its properties after some time. Utilizing AAC can lessen development costs by about 2.5 percent for structures, for example, schools and medical clinics, and diminish the running expenses of lodging and places of business by 30 to 40 percent after some time. Figure 1 shows an AAC block, (brand name as Aerocon blocks).

Figure 1. Aerated Autoclaved Concrete (AAC) blocks

The Autoclaved aerated concrete (AAC) was created and developed in 1924 by the Swedish architect Dr. Johan Axel Eriksson, along working with Professor Henrik Kreüger at the Royal Institute of Technology [1]. It is an eco-accommodating structural material that originates from modern waste and is produced by using non-poisonous and non-toxic ingredients. With AAC, the development procedure can be around 20 percent quicker. It weighs just around 50 percent of a standard solid square and has high warm protection and is acoustics-accommodating. It additionally has preferable imperviousness to fire over fly debris and is non-burnable. It's non-hypersensitive and henceforth keeps up the nature of air inside a structure without changing its properties after some time. As indicated by one report, AAC now represents over 40% of all development in the United Kingdom and over 60% of development in Germany [2].

2.   Raw Materials Used in Manufacturing of AAC Blocks

As compared to most other concrete building applications in the construction industry, Aerated Autoclaved Concrete (AAC) is created utilizing no total bigger than sand. Quartz sand, calcined gypsum, lime (mineral) as well as concrete, and water are utilized as coupling specialists. In certain nations, similar to India and China, fly debris produced from thermal power plants and having 50-65% silica content is used as an aggregate. Many raw materials are used in the manufacturing of AAC blocks. They are summarized in Table 1.

Table 1. Percent proportion of raw materials

S. No.	Material	% of proportion for AAC Blocks with Fly ash / with Sand
1	Flyash / Sand	65-70 / 55-65 %
2	Cement - 53 GRADE OPC	6-15 / 10-20 %
3	Lime	18-25 / 20-30 %
4	Gypsum	$3 - 5 / 2 - 3$ %
5	Aluminium powder paste (600 kg / m^3)	8 %
6	Water	$0.6 - 0.65$ %
7	Casting system	36-40 ° C / 35-38 ° C

3.   Specifications and Performance Summary of AAC Blocks

The product specification and its performance summary of AAC blocks are summarized as below [3]:

3.1   Appearance

The Autoclaved aerated concrete (AAC) is light-hued and contains numerous small voids that can be observed when taken a gander at intently. The gas used to 'froth' the solid during the manufacturing process is hydrogen from the chemical reaction from the Aluminum paste with alkaline soluble components in the cement concrete. These air pockets add to the material's protecting properties. In contrast to stonework, there is no immediate way for water to go through the material; be that as it may, it can suck up dampness and a suitable covering is required to forestall water infiltration.

3.2   Size and Density

The Autoclaved aerated concrete (AAC) blocks are made of 625 mm length, 250 mm tallness, and of different thicknesses: 100, 125, 150, 200, 225, 250, 300 mm with a resistance of ±1.5 mm. The thickness of the block is 600 to 650 kg/cum, though the blocks have a thickness of 1750 to 2000 kg/cum. The density of wet blocks is around 800 kg/cum as compared to that of red clay bricks (2400 kg/cum).

3.3   Structural Capability

The compressive quality of Autoclaved aerated concrete (AAC) blocks is excellent. The compressive quality is from 35 to 50 kg/cm^2 (according to IS: 2185). Although it is one-fifth the thickness of typical solid it despite everything has a large portion of the bearing quality, and load bearing structures up to three stories high can be securely raised with AAC blockwork. The AAC is 3-4 times lighter than traditional bricks, therefore, easier and cheaper to transport. Usage reduces the overall dead load of a building, thereby allowing the construction of taller buildings. Entire building structures can be made in AAC from walls to floors and roofing with reinforced lintels, blocks, and floor, wall, and roofing panels available from the manufacturer. AAC floor panels can be used to make non-load bearing concrete floors that can be installed by carpenters. Lightweight blocks diminish the mass of a structure, along these lines diminishing the effect of tremor on a structure.

3.4   Thermal Mass

The thermal mass performance of Autoclaved aerated concrete (AAC) is dependent on the climate in which it is used. With its mixture of concrete and air pockets, AAC has a moderate overall level of thermal mass performance. Its use for internal walls and flooring can provide significant thermal mass. The temperature moderating thermal mass is most useful in climates with high cooling needs.

3.5   Insulation

The Autoclaved aerated concrete (AAC) has very good thermal insulation qualities relative to other masonry. A 200mm thick AAC wall gives an R-value rating of 1.43 with 5% moisture content by weight. With 2–3mm texture coating and 10mm plasterboard internal lining it achieves an R rating of 1.75 (a cavity brick wall achieves 0.82). A texture-coated 100mm AAC veneer on a lightweight 70mm or 90mm frame filled with bulk insulation achieves a higher R rating than an otherwise equivalent brick veneer wall (see Insulation; Lightweight framing). Relative to their thickness, The Thermal Conductivity of AAC blocks is 0.16 kW/m°C against 0.70 bricks, thus recurring energy cost is reduced in air conditioning. AAC panels provide less insulation than AAC blockwork, e.g. a 100mm block work AAC wall has a dry state R-value of 0.86 and a 100mm AAC wall panel has a dry state R-value of 0.68.

3.6   Sound Insulation

With its closed air pockets, Autoclaved aerated concrete (AAC) can provide very good sound insulation. It has superior sound absorption qualities due to the porous structure of blocks. Combining the AAC wall with an insulated asymmetric air cavity system gives a wall excellent sound insulation property. AAC offers sound attenuation of about 42 dB, blocking out all major sounds and disturbances. The Sound Reduction Index is 45 dB for 200 mm thick block walls (against 50 for 230 mm thick wall). It is ideal for schools, hospitals, hotels, offices, multi-family housing, and other structures that require acoustic insulation.

3.7   Fire and Vermin Resistance

Autoclaved aerated concrete (AAC) is inorganic, incombustible, and does not explode; it is thus well suited for fire-rated applications. Depending on the application and the thickness of the blocks or panels, fire ratings of up to four hours can be achieved. AAC is non-combustible and fire-resistant up to 1600° C. It can withstand up to 6 hours of direct exposure. Due to the structure of blocks, AAC cannot be damaged or infested by termites and other pests. It does not attract rodents or other pests nor can it be damaged by such.

3.8   Durability and Moisture Resistance

The purposely lightweight nature of Autoclaved aerated concrete (AAC) makes it prone to impact damage. With the surface protected to resist moisture penetration, it is not affected by harsh climatic conditions and does not degrade under normal atmospheric conditions. The level of maintenance required by the material varies with the type of finish applied. The porous nature of AAC can allow moisture to penetrate to depth but the appropriate design (damp proof course layers and appropriate coating systems) prevents this from happening. AAC does not easily degrade structurally when exposed to moisture, but its thermal performance may suffer. Several proprietary finishes (including acrylic polymer-based texture coatings) give durable and water-resistant coatings to AAC blockwork and panels. They need to be treated similarly with acrylic polymer-based coatings before tiling in wet areas such as showers.

3.9   Water Absorption

In Autoclaved aerated concrete (AAC) curing takes place at a high temperature and high pressure in saturated steam. During curing, part of the siliceous material (flyash) reacts chemically with the calcareous ingredients such as lime liberated by the hydration of cement to form a Micro-Crystalline structure of Tobermorite with a much lower specific surface and is characterized by pores formed by the release of H_2 gas during Casting-Rising stage of production.

3.10   Toxicity and Breathability

The aerated nature of AAC facilitates breathability. There are no toxic substances and no odor in the final product. However, AAC is a concrete product and calls for precautions similar to those for handling and cutting concrete products. It is advisable to wear personal protective equipment such as gloves, eyewear, and respiratory masks during cutting, due to the fine dust produced by concrete products. If low-toxic, vapor-permeable coatings are used on the walls, and care is taken not to trap moisture where it can condense, AAC may be an ideal material for homes for the chemically sensitive.

3.11   Constructability, Availability and Cost

Although AAC is relatively easy to work, it is one-fifth the weight of concrete comes in a variety of sizes and is easily carved, cut, and sculpted, it nevertheless requires careful and accurate placement: skilled trades and good supervision are essential. Different sizes of blocks help reduce the number of joints in wall masonry. Lighter blocks make construction easier and faster. It reduces construction time by 20%. They are easy to install. AAC sets and hardens quickly. Blocks can be easily cut, drilled, nailed, milled, and grooved to fit individual requirements. Thick-bed mortar is more forgiving but is uncommon and not the industry preferred option. It also simplifies hydro-sanitary and electrical installations, such as pipes or ducts, which can be installed after the main construction is complete. The construction process with AAC produces little waste as blockwork off-cuts can be reused in wall construction.

4.   Autoclaved Aerated Concrete (AAC) Block Walling System

The AAC blocks are generally used in wall masonry. The process of laying and construction of a complete wall unit is summarized in the following steps [4].

4.1   Laying of Blocks

In laying AAC blocks, the procedure is the same as that for conventional half brickwork except that the block should be slightly wet with a sprinkler before use and not soaked as in the case of bricks.

a) Before laying the first course the alignment of the wall is marked over the DPC.

b) The blocks for 1st course should first be laid dry without mortar along a stretched thread between properly located corners of the wall to determine the correct position of the blocks, including those of cross walls joining it, and also to adjust their spacing within the wall length.

c) When the blocks are set in a proper position, the two corner blocks are removed. A specified mortar bed is spread for the required bed thickness (10mm).

d) The blocks are laid back in place with true level and plumb.

e) The thread is then stretched tightly along with the faces of the two corner blocks and the faces of the intermediate blocks are adjusted to coincide with the thread line.

f) Each intermediate block is removed and re-laid with mortar.

g) After every three or four blocks, layers are laid, their correct alignment level and verticality are checked.

h) In the vertical joints, the mortar is applied only on the face of the blocks.

4.2 Electrical and Plumbing Installations

Electrical and plumbing installations in AAC masonry are placed in routed chases. Care should be taken when laying out chases to ensure that the structural integrity of the AAC elements is maintained. Do not cut reinforcing steel or reduce the structural thickness of the AAC elements except where permitted by the designer. In vertically spanning AAC elements, horizontal routing should be permitted only in areas with low flexural and compressive stresses. In horizontally spanning AAC elements, vertical routing should be minimized.

4.3 Exterior Finishes

Unprotected exterior AAC deteriorates when exposed to cycles of freezing and thawing while saturated. To prevent such freeze-thaw deterioration, and to enhance the aesthetics and abrasion resistance to AAC, exterior finishes should be used. They should be compatible with the underlying AAC in terms of thermal expansion and modulus of elasticity and should be vapor permeable. Many different types of exterior finishes are available. Polymer-modified stuccos, paints, or finish systems are the most common exterior finish for AAC. They increase the AAC's water-penetration resistance while allowing the passage of water vapor. Heavy acrylic-based paints containing aggregates are also used to increase abrasion resistance. There is generally no need to level the surface, and horizontal and vertical joints may be chamfered as an architectural feature or may be filled.

4.4 Interior Finishes

Interior finishes are used to enhance the aesthetics and durability of AAC. They should be compatible with the underlying AAC in terms of thermal expansion and modulus of elasticity and should be vapor permeable.

a) Many different types of interior finishes are available. Interior AAC wall panels may have a thin coat of mineral-based plaster to achieve a smooth finished surface. Lightweight interior gypsum-based plaster may provide a thicker coating to level and straighten walls, and to provide a base for decorative interior paints or wall finishes. Interior plasters have bonding agents to enhance their adhesion and flexibility and are commonly installed by either spraying or troweling.

b) For commercial applications requiring high durability and low maintenance, acrylic-based coatings are often used. Some contain aggregates to enhance abrasion resistance.

c) When ceramic wall tile is to be applied over AAC, surface preparation is normally necessary only when the AAC surface requires leveling. In such cases, a Portland cement or gypsum-based parge coat is applied to the AAC surface before setting the ceramic tile. The ceramic tile should then be adhered to the parged wall using either a cement-based thin-set mortar or an organic adhesive. In moist areas such as showers, only a Portland cement-based parge coat should be used, and the ceramic tile should be set with cement-based, thin-set mortar only.

4.5   Service Lines

For concealed or piping, a block wall can be chased using a hand or electric router. Depths of vertical chases should be limited to one-third of the wall thickness and horizontal chases to one-sixth of the wall thickness. Holes in a block wall can be made with a standard hand or electric drill. The chases shall be refilled with a leaner mortar and chicken mesh shall be applied to that area and cured (Figure 2).

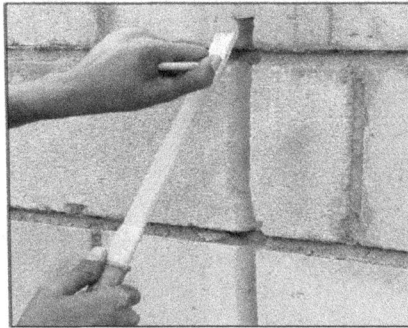

Figure 2. The chasing of the wall for service lines

4.6   Lintels

Precast or cast-in-situ concrete lintels can be used in block masonry, over all openings. Lintels shall always rest on a full block with a minimum bearing as under. Below the openings, the RCC band should be provided with reinforcement to avoid diagonal tension cracks. The bond beam to be extended up to 300 mm from window corners on both sides. Table 2 shows the minimum bearing on each side for the different opening sizes.

Table 2. Minimum bearing on each side for different opening sizes.

Opening Size	Upto 900 mm	900 to 2000 mm	2000 to 3000 mm	Above 3000 mm
Minimum bearing (each side)	150mm	200mm	300mm	To design

4.7   Plastering

Followings are the points that should be considered while plastering the AAC walls:

a) Do not soak the wall before plastering. The wall shall be moistened evenly before applying the plaster. A fog spray is recommended for this purpose.

b) For external plastering has to be carried out in two coats, apply SBR coating with sand on the block surface will enhance the bonding and minimize the thickness of plastering.

c) It is recommended to use cement mortar 1:6 for internal & external plastering works and preferably use PPC cement for masonry and plastering works to minimize shrinkage cracks.

d) Plastering thickness can be minimized to 10 mm and 15 mm for internal and external walls.

4.8 Precautions while Laying the AAC Masonry Blocks

The cracks occurring in block masonry and plastering are of any structural problem involving stability and safety to the structure. But it is advised to minimize the same to have a good appearance and maintenance-free.

a) Do not store the blocks on an unleveled surface

b) Do not use wet blocks for masonry construction

c) Do not make the holes on block masonry for scaffolding supports

d) Do not soak the blocks before use

e) Do not hammer the block masonry for service lines, chases, etc.

f) Do not completely wet the block masonry before plastering works

g) Do not chase the blocks back to back for lesser thickness blocks.

5. Advantages of Using AAC Block

Autoclaved aerated concrete (AAC) has many advantages as compared to other cement concrete materials. The basic advantage is that it is cost-efficient and Eco-friendly with having a low environmental impact. In the manufacturing of AAC blocks, the topsoil of the earth's surface is not used, therefore it emits very low carbon dioxide as compared to red clay bricks. Since AAC blocks are an industrial product manufactured with machines, the quality of the end products is very good, uniform, and consistent. The blocks are even and well finished from all sides, therefore the thickness of wall plaster is reduced. Due to its lightweight, there will also be a reduction of dead weight in the structural system, which will lead to the saving of steel and concrete. Also, this leads to the construction of more stories or taller buildings. The economy is achieved in multistoried, especially buildings by using AAC blocks as compared to red clay bricks due to a significant reduction in the dead weight of the structure; and hence it reduces the cost of the RCC framed structure.

The AAC blocks have a lot of pores and voids; hence it provides better sound absorption and insulation as compared to red clay bricks or concrete blocks. The AAC blocks have a low thermal conductivity which is approximately 0.24 kW-M/°C, which results in the saving of electricity costs by 30%, which is required for heating and cooling of the house. The manufacturing process of AAC blocks is non-polluting. The by-product from the manufacturing industry is only steam. All the ingredients to manufacture the AAC blocks are non-toxic and safe. The AAC blocks are fire-resistant and non-combustible. They can tolerate up to 6 hours of direct exposure to fire. The AAC blocks have air voids and hence have better fire-resisting properties compared to red clay bricks. The melting point of the AAC blocks is over 1600 °C more than twice the typical temperature in building fire 650 °C. The AAC blocks are easy to work with i.e. they can be easily cut nailed and drilled and can be fitted to the individual requirements which give more design flexibility. It cannot be rotten easily since it is pest and termite-proof. The AAC walling system has simplified hydro-sanitary and electrical installations, such as pipes or ducts, which can be installed after the main construction is completed. It also has low maintenance and it reduces the operating and maintenance cost by 30% to 40%. The chemical mortars can be used for joining the AAC blocks in masonry work; hence this reduces the material consumption for cement mortar and also there is no need for curing. The AAC blocks are suitable both for non-load bearing strictures and also for reinforced cement concrete structures in partition walls.

6. Disadvantages of AAC Block

There are many advantages of using AAC block, which has made this material a very suitable and sustainable material in the building and construction sector. But there are also a few disadvantages of using AAC blocks. It has been observed that the Aircrete cracks after installation in the rainy season, which can be avoided by reducing the strength of the mortar and ensuring the blocks, are dry during and after installation. The AAC blocks should be handled carefully then as compared to red clay bricks to avoid breakages since it is little brittle. Due to its brittle nature, it requires long thin screws when fitting cabinets and wall hangings. The initial cost of the manufacturing industry is a bit high. There are not many factories, which are producing AAC blocks; therefore it is not very easily available.

7. Comparative Analysis of AAC Block, Clay Brick, and CLC Block

The comparative analysis of Autoclaved Aerated Concrete (AAC) blocks, Red clay bricks, and Cellular Light Weight Concrete (CLC) blocks are summarized in Table 3:

Table 3. Comparison of AAC Block, Clay Bricks, and CLC Blocks [5].

Parameter	AAC Blocks	Clay Red Bricks	CLC Blocks
Raw Materials	Cement, fly ash, water, and Air entraining agents	Locally available clay	Cement, lime, especially ground sand, foam
Size	400-600mm X 200mm X 150mm – 300mm	225mm X 75mm X 100/150mm	400-600 x 200 x 100/150/200 mm
Variation Size	1.5 mm (+/-)	5 mm (+/-)	5 mm (+/-)
Compressive Strength (As per IS codes)	3-4 N/mm2	3.5 N/mm2	2 -2.5 kg/cm2
Dry Density (As per IS codes)	550-650 kg/m3 Its one-third of the weight of clay brick which makes it easy to lift and transport	1800 kg/m3	800 kg/m3
Cost-Benefit	For high-rise buildings, there is the reduction of Deadweight which leads to saving in Concrete and steel.	As easily available in the local market hence it is beneficial for low rise structure.	For high-rise buildings, there will be a reduction of Deadweight which leads to saving in Concrete and steel quantities.
Fire Resistance (8″ Wall)	Up to 4 Hours	Around 2 Hours	Around 4 Hours
Quality of End Product	Factory-made product. So the quality of the end product is consistent and good	Locally made products. Quality depends on various parameters like raw materials quality, the manufacturing process,etc.,	The quality of the end product depends on the foam used and the degree of quality control
Sound Insulation	Better Sound absorption /insulation as compared to bricks	Normal	Better Sound absorption /insulation as compared to bricks
Energy Saving	Low thermal conductivity (0.24 Kw-M/C) helps in saving electricity costs 30% for heating and cooling.	High thermal Conductivity (0.81 Kw-M/C). So no significant cost savings	Low thermal conductivity (0.32 Kw-M/C) helps in saving electricity costs 30% for heating and cooling.

Environmental Friendliness	In AAC Block there is no topsoil consumption and it emits very low CO_2 as compared to Red clay bricks while manufacturing	One sq. ft of carpet area with clay brick walling will consume 25.5 kg of topsoil (approx.). It damages the environment	In CLC Block there is no topsoil consumption and it emits very low CO_2 as compared to Red clay bricks while manufacturing.
Internal and External Plaster	As these bricks have dimensional accuracy, the internal and external plaster thickness is reduced	Requires thick plaster surface as there are variations in the dimensions	As these bricks have dimensional accuracy, the internal and external plaster thickness is reduced
Cost of Construction (year 2018)	1 Cum costs – Rs. 4200/-	1 Cum costs – Rs. 2440/-	1 Cum costs – Rs. 4000/-
Joining Process	Chemical mortars can be used for joining the brick. This reduces the material consumption for cement and avoids the curing process	The traditional mortar needs to be used and the brickwork should be cured at least for 7 days before plastering	Chemical mortar can be used for joining the brick. This reduces the material consumption of cement & avoids the curing process.
Availability	The factory setup cost is high. Not many factories, so availability is a concern.	Available locally in all cities and villages.	Factory setup cost is low as compared to AAC. It also takes a long time to produce if steam curing is not used. Timely availability is a concern.
Thermal Insulator	AAC Blocks are good thermal insulator, monthly expenses it will save cost for an entire lifetime	It has low thermal insulation as compared to AAC and CLC Block	CLC Blocks are good thermal insulators. monthly expenses it will save cost for an entire lifetime
Tax Contribution	Contributes to Government taxes in form of Central, Excise, and VAT	No Tax Contribution	Contributes to Government taxes in form of Central Excise and VAT
Cylindrical Structures	For Cylindrical structure, these blocks are not much useful	Cylindrical manholes need small size of bricks so that the curvature can be formed hence Red clay bricks are useful	For Cylindrical structure, these blocks are not much useful

| Water Absorption | Absorb 12- 15% by the total volume of AAC blocks | Absorb 17 -20% by the total volume of red clay brick | Absorb 12-15% of water by the total volume of Block |
| Range of Application | They are suitable for non-load-bearing or RCC structure in the partition wall | They are useful in both load-bearing and non-load bearing structure | They are suitable for non-load-bearing or RCC structure in the partition wall |

8. Cost Comparative Analysis of AAC Block, Clay Brick, and CLC Block

The cost comparative analysis of brickwork in masonry and plaster for AAC blocks, Red clay bricks, and Cellular Light Weight Concrete (CLC) blocks are summarized in Table 3 and Table 4 respectively [6].

Table 3. Cost comparative analysis for AAC blocks and clay brick masonry for 1M³ [1:4]

Parameters	Clay Red Bricks	AAC Blocks
Quantity Analysis	200mmx 100mm x 100mm	600mm x 200mm x 200mm
No. of bricks / blocks	500	37
Mortar Quantity	0.2766 M³	0.1344 M³
No. of begs of cement	1.65	1.0
Quantity of Sand	0.221M³	0.1075 M³
Quantity of Water	31 Litres	16 Litres
Rate Analysis	5252.00 Rs./ m² (As per MP PWD SOR building work 2014 clause no.6.3)	5052.00 Rs. / m² (As per MP PWD SOR building work 2014 clause no.6.27)

Table 4. Cost comparison for plasterwork for AAC blocks and clay brick for 1 M³ [1:4]

Parameters	Clay Red Bricks	AAC Blocks
The volume of mortar for plaster	1.8 M³	1.0 M³
The volume of mortar by 25% for wastage and frog filling	2.25 M³	1.25 M³
Quantity of cement	0.45 M³	0.25 M³
No. of begs of cement	13.5	7.5
Quantity of Sand	1.8 M³	1.0 M³
Quantity of Water	236.25 Liters	131.25 Liters
Rate Analysis	171.00 Rs./ m² (As per MP PWD SOR building work 2014 clause no.13.6)	91.10.00 Rs./ m² (As per MESSOR building work 2010 item no.14001)

9. Environmental Benefits of AAC Blocks

The Autoclaved Aerated Concrete (AAC) is an eco-friendly material that has many environmental benefits. The weight of the AAC block is around one-fourth to one-fifth that of concrete based on volume. The manufacturing of AAC blocks has the same greenhouse gas environmental impact and has the same embodied energy as that of concrete blocks. The AAC blocks or panels have lower embodied energy per square meter than a concrete alternative building material. The AAC block and panels have more insulation value and thus it has low energy usage for heating and cooling loads requirement. The total energy used in manufacturing the ACC blocks is around 50% less than that of manufacturing other prefabricated building components and products. As compared to regular cement concrete building products, AAC reduces around one-third of the environmental waste. The Autoclaved Aerated Concrete (AAC) blocks and panels have proven to be more durable, provide thermal insulation and structural requirements, and also have major economic and environmental benefits as compared to other traditional building components and products. Thus Autoclaved aerated concrete can be said to a suitable and potential eco-friendly building material, which is beneficial for the environment, which fulfills the requirement for the construction of sustainable architecture and construction [7].

Conclusions

The Autoclaved Aerated Concrete (AAC) is a novel and one of the most suitable and sustainable building materials in the present building construction industry. AAC blocks are a result of productive use of recycled industrial waste i.e. fly ash, hence this material can be classified as a sustainable building material. The production price of AAC blocks at the manufacturing unit is Rs. 3200/- to Rs. 3600/ per cubic meter as per the rates in India in the year 2019. The inherent properties of AAC blocks result in fast and efficient construction techniques. Hence the Autoclaved Aerated Concrete (AAC) has become an efficient building construction material that is being used in a wide range of residential, commercial, and industrial buildings and it has been used in the Gulf countries for the last 40 years and in Europe for since 70 years, and in Australia and South America for the past 20 years. According to a report, the AAC blocks are used in more than 60% of construction in Germany, and England approximately 40% of all construction industry [8]. Since the AAC blocks use readily available raw materials in the manufacturing process, have excellent durability, are energy efficient, are cost-effective, and also can be recycled, therefore Autoclaved Aerated Concrete (AAC) can be said to be a green and sustainable building material.

References

[1]   Five green building blocks, information on
       http://www.thehindu.com/features/homes-and-gardens/5-green-building-
       blocks/article4813910.ece

[2]   S. Schnitzler, Autoclaved Aerated Concrete as a Green Building Material, UC Davis Extension, Switzerland, 2016.

[3]   M. Arif Kamal, Autoclaved Aerated Concrete (AAC): A Sustainable Building Material information on http://www.masterbuilder.co.in/autoclaved-aerated-concrete-aaca-sustainable-building-material/

[4]   New Building Materials and Technologies, Vol. IV, Compendium of New Building Technologies, Indian Building Congress, New Delhi, India, 2019.

[5]   Comparison of AAC Blocks vs CLC Blocks vs Red Clay Bricks, information on https://happho.com/comparision-aac-blocks-vs-clc-blocks-vs-red-clay-bricks/

[6]   U. Jain, M. Jain, S. Mandaokar, Comparative Study of AAC Blocks and Clay Brick and Costing, International Journal of Research in Engineering, Science and Management Vol. 1, Issue 9, 2018.

[7]   Global Autoclaved Aerated Concrete Market Outlook: Trend and Opportunity Analysis, Competitive Insights, Actionable Segmentation and Forecast 2023, Research Report, Energias Market Research, 2019, Buffalo, USA.

[8]   Autoclaved Aerated Concrete (AAC) A Sustainable Building Material, to Witness a CAGR of 7.9% during 2017 – 2023, information on https://www.globenewswire.com/news-release/2018/02/23/1386502/0/en/Autoclaved-Aerated-Concrete-AAC-A-Sustainable-Building-Material-to-Witness-a-CAGR-of-8-0-during-2018-2024-Energias-Market-Research-Pvt-Ltd.html

Novel Approaches in Materials and Technologies for Building Construction   Materials Research Forum LLC
Materials Research Foundations **158** (2024)   https://doi.org/10.21741/9781644902936

CHAPTER 3

Hydraform Interlocking Block Walling System

1.   Introduction

The conventional brick walls, stone walls and even hollow concrete blocks require mortar to hold the individual building units in desired position. A new concept of concrete block construction without mortar has been developed, but due to lack of proper promotion and lack of confidence as to their structural stability, they have not been largely used in India (Figure 1). Many such ideas of different type and shape of blocks for jointing without cement or mortar have been developed and used in many countries. The basic theory behind elimination of mortar is to devise an interlocking system of blocks which are technically sound, structurally safe for vertical and wind loads. Besides the cost of mortar itself the time spent in making and placing mortar can be reduced to lay up the blocks.

Figure 1. A Hydraform Interlocking block

2.   Interlocking Application of Hydraform Blocks

Interlocking in wall masonry is very unique and one such Technology is known as Hydraform Interlocking Building system that replaces the conventional brick and mortar by using Hydraform machine made interlocking blocks. The other components of the conventional building system remain largely unchanged. The application can be as dry stacked or with cement slurry in Interlocking tongue and groove that enables speedier construction of high quality aesthetic and affordable building in stretcher bond as well as in the normal English/Flemish bond with mortar. The Blocks have an extremely appealing face-brick finish due to sharp edges. The walls may be left exposed, plastered or finished with cement paint.

This technology has been in use in a number of countries for almost 25 years as a load bearing option using the compressed stabilized earth blocks. In Indian subcontinent interlocking technology option has been put to use for the last few years. The versatility of technology has also been validated by use in construction projects by Housing and Urban Development Corporation (HUDCO) and Building Materials and Technology

Promotion Council (BMTPC) in some of the projects promoting newer and sustainable building technology options. The self-alignment of interlocking blocks, without mortar results in very rapid construction, when compared to conventional block construction. To get high strength of the blocks (for load bearing walls for 4 to 5 storeys) the blocks are moulded under high pressure using a precision system rather than the usual casting process. The blocks can be cast mechanically but production is relatively slow.

Interlocking Technology is dry stacking mortar - less method of constructing walls. Block are not laid with mortar, they rely on the interlocking mechanism to provide resistance to applied loads. Hydraform interlocking dry stacking utilizes interlocking mechanisms of shear keys as well as self-weight to resist the external loads. Dry stacking results in reduction of construction costs due to saving in construction time, reduced requirement for skilled labour and costly material especially cement and reusability of the blocks. The requirement for unskilled labour makes dry stacking particularly attractive to labour makes dry stacking particularly attractive to labour based work [1]. As per the requirement of IS 4326:1993, a thin mortar of the specified type can be used even in these Interlocking types of the blocks. Interlocking block masonry has also been validated by Gujarat State Disaster Management Authority for earthquake resistant construction even in zone V with appropriate structural bands.

3.   Hydraform Interlocking Blocks

   a) The blocks are mainly of following size and dimension to suit standard application requirements. However size can be tailored for large quantity application requirements [2].

Figure 2(a), (b) and (c) shows the size and dimensions of interlocking blocks.

Figure 2(a) –HF: 220Figure 2(b)–HF: 220mm Conduit Figure 2(c) –HF: 150 mm

b) The Following sizes are available in Hydraform Interlocking Blocks:

Types of blocks	Width	Length	Height	Weight (kg)
220	220 mm	100-240 mm	115 mm	9 – 11 (full size)
200	190 mm	240 mm	115 mm	8 – 9 (full size)
115 (Partition walls)	115 mm	100-220 mm	115 mm	4.5 – 6 (full size)

Figure 3(a), (b), (c) & (d) shows Isometric view, Side Elevation, End Elevation and Plan of an concrete interlocking block.

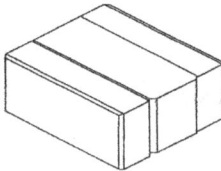

Figure 3 (a) – Isometric View

Figure3 (b) – Side Elevation

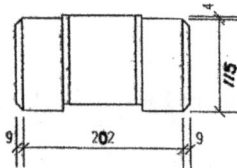

Figure3 (c) – End Elevation

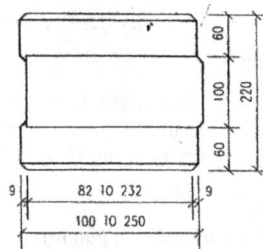

Figure3 (d) - Plan

4. Hydraform Block Making Machine

Hydraform Blocks are manufactured by high-pressure hydraulic compaction of the mixed material in a Hydraform Block making machine. Model of various range of interlocking Block making machines as per specification are available. The machine comprises of an engine or Electric motor, hydraulic power pack including cylinders, compression chamber, pre-compacting and double compaction, loading assembly on static frame are all fitted on a roadworthy frame inclusive of spring-axle, tow-hitch and road tyres (Figure 4). Hydraform has range of block making machines and mixing options to suit project requirement and location.

Figure 4. Hydraform block making machine with integrated pan mixer (Code M7 MI)

5. Mechanized Production of Blocks

5.1 Mix

The mix design indicating the proportion of various ingredients is worked out carefully keeping in view the target dry density of the block.

5.2 Batching

The quantities of various ingredients are proportioned on the basis of weight, with due correction being made to the quantities on account of inherent moisture content of other ingredients and quantity of water in the Fly Ash and sand.

5.3 Mixing

The ingredients are mixed in a pan mixer having preferably rollers & scrapers for mixing. Cleaned mixer is started before charging the materials, to obtain optimum performance, Sand and fly ash with other ingredients is first fed into the mixer, and thoroughly mixed to ensure even distribution of cement. The appropriate amount of water is slowly added thereafter, keeping on mixing till homogenous colour is achieved.

5.4 Formation of Block in the Hydraform Machine

The mix is then fed in the Hydraform Blocks making machine moulding chamber through the hopper and hydraulic pressure is applied to make a block. The block to be extruded and placed at even grounded Stack yard.

Materials Research Forum LLC
https://doi.org/10.21741/9781644902936

5.5 Curing of Blocks

The green blocks are immediately covered with Polythene sheet to protect moisture within the block. The blocks are sprinkled water cured at least twice a day for 14-21 days. The curing period shall depend upon contents of cement / lime and weather.

6. Classification of Blocks

Different classes of blocks can be made as per compressive strength requirement in accordance with:

(i) SEB (Soil Earth Stabilized) Blocks in accordance with IS 1725-1982.

(ii) FAL-G(Flyash - lime) Blocks in accordance with IS 12984-1984.

7. Raw Materials Used in Manufacturing

The main raw materials for production of Hydraform interlocking blocks are Soil and Flyash, which are available cheaply and in abundance. Using SEB & FAL-G blocks is a low cost, attractive and eco-friendly option available for constructing dwellings. Both soil and flyash are stabilized by various stabilizing agents such as cement, lime, gypsum etc. and then compressed and compacted in the Hydraform block machine to give the finished product [2].

These blocks are made following respective Codes prescribed for different Raw Materials viz

IS 1725:1982 (Specification for soil-based blocks used in general building construction)

IS 12984: 1990 (Flyash- lime bricks- specification?)

8. Properties of Hydraform Interlocking Blocks

8.1 Compressive Strength

The Hydraform Interlocking blocks/bricks when tested in accordance with IS: 3495 Part I-1976 should have a minimum compressive strength after 28 days of curing us follows.

Hydraform 'SEB' Blocks

	Class designation	Compressive strength in (Kg/cm^2)
i)	30	not less than 30
ii)	75	not less than 75

Hydraform Fly Ash Blocks (FAL-G)

	Class designation	Compressive strength in (Kg/cm^2)
i)	75	not less than 75
ii)	100	not less than 100
iii)	125	not less than 125

8.2   Test Procedure

(i) The compressive strength of the blocks can be easily tested at the site by using the Hydraform blocks testing Machine.

(ii) Compressive strength test is done in compression testing machine. Block is placed between the jaws and load is applied gradually. Precaution is taken such that load is applied to the flanged portion of the blocks. For this two steel plates of sizes 50mm x 240mm and thickness 10 mm are placed on top flange and gradual load is applied over the plates till the failure occurs and not the maximum load at failure (Figure 5).

The test result shall be calculated as given below:

$$\text{Compressive strength} = \frac{\text{Maximum load at failure}}{\text{Average net area of flanged portion}}$$

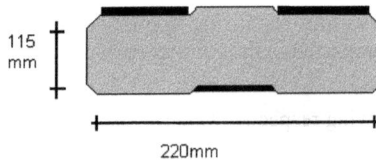

Figure 5. Steel Plates placed on the flange for performing compressive test.

8.3   Water Absorption

The Hydraform Interlocking Blocks when tested in accordance with the procedure laid down in IS 3495 Part II-1976, after immersion in water for 24 hours, an average absorption should not be:

a) For the blocks/bricks with (FAL-G (Flyash/Cement) Bricks/ Blocks) more than 15% (by weight).

b) For the blocks/bricks with (SEB (Soil Earth Stabilized) Bricks/ Blocks) more than 15% (by weight).

8.4   Drying shrinkage

The average drying shrinkage of the blocks when tested by the method described in IS 4139:1989 shall not exceed 0.15%.

8.5   Weathering

When tested in accordance with IS 1725-1982, Appendix A, the maximum loss of weight shall not exceed 5%.

9.   Construction Process of Hydraform Block Masonry

A dry stacked interlocking building is laid on conventional strip footing. Foundation walls are built with blocks of higher strength laid in mortar bed or even conventional type foundation. The courses are laid on mortar up to the floor level. Courses above floor level are dry stacked up to lintel band. The top three courses are held together by mortar to form a ring beam at the top. The alternative is to cast a reinforced concrete ring beam. In areas with low technical skills, it is preferred to use bricks force and mortar option. The roof is tied to the ring beam to prevent uplift. Different conventional finishes can be applied to suit the aesthetic needs of the owner.

9.1   Interlocking Mechanism of Hydraform Blocks

The locking of a male face of one block with the female face of another or the locking of the bed of one block with the ridge of the one below it is called 'Interlock' (Figure 6).

Figure 6. The interlocking mechanism of masonry blocks

9.2   Bed and Ridge

The recessed under surface of the block is referred to as the bed. The raised top surface of the block is called the Ridge (Figure 7).

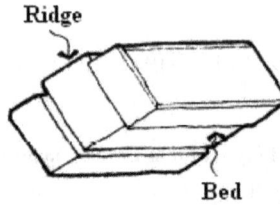

Figure 7. The bed and ridge of the interlocking block.

9.3   Course

One (horizontal) layer of Hydraform blocks is called a course.

Height of a course = 115 mm (with a new set of wear plates). Figure 8 shows the odd and even courses of the masonry.

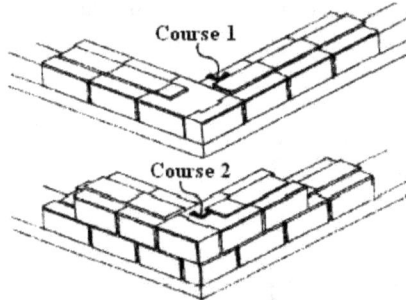

Figure 8. The odd and even course of the interlocking block masonry.

9.4   Corners

For a corner, shaved full blocks or shaved ½ blocks are required. Also for the corner block shave off the ridge and male face of the corner block, as shown in Figure 9. Ensure that the shaved ridge points upward and the shaved male face point's outwards. One must start the first course with a ½ block.

Materials Research Forum LLC

https://doi.org/10.21741/9781644902936

Figure 9. The corner masonry using interlocking blocks

10. Compatibility with Earth Quake Resistant Construction

The Hydraform interlocking blocks can be easily reinforced (because of the grooves) against the conventional masonry. All the relevant bands i.e. roof bands plinth band, lintel band, gable band etc. can be easily be incorporated in the masonry. Both vertical and horizontal reinforcement can be provided by means of the grooves. For use in earthquake prone areas and their specifications are given in the GSDMA guidelines. (Prepared by Gujarat State Disaster Management Authority)

11. Load Bearing Masonry

Since blocks are of 220 mm width and can be made of strength > 75 kg/sq.cm, same can be safely used for Load Bearing construction. Depending on structural requirements of the building, appropriate lintel & roof bands can be used. Hydraform Company has done tests from time to time for conformity of dry stacked masonry in G+2 storey buildings. Fly Ash based interlocking blocks can be made of higher compressive strength to suit the Load bearing construction requirements beyond Ground floor to suit structural requirements. In terms of IS 1905, masonry of M7 grade requires use of brick / block to be of M7 grade and mortar of class I. Use of Interlocking blocks of M7 grade with thin mortar slurry of 1:3 (1 cement: 3 fine sand), if required, will satisfy this requirement. Figure 10 and Figure 11 shows the use of Hydraform interlocking blocks in mid-rise buildings in the state of Gujarat, India.

Figure 10. A Four storeyed housing with interlocking blocks at Vapi, Gujarat, India

Figure 11. A Three storeyed housing with interlocking blocks at Anjar, Gujarat, India

12. Block Masonry in Framed Structures

Framed construction mainly require brick/block work to be used as an infill only, therefore dry stacked interlocking block work can be used in walls of +- 220 mm, 190 mm thickness. For block work of lesser width it is recommended to use 1:3 mix mortar slurry. The blocks have standard height of 115 mm which makes it easier to design the beam height for required number of courses. Ends with beams and columns are same as in brick. Figure 12 shows the use of interlocking blocks in multi-storeyed housing.

Materials Research Forum LLC
https://doi.org/10.21741/9781644902936

Figure 12. A multi-storeyed housing with interlocking blocks

13. Reinforced Masonry

Interlocking Blocks with horizontal and vertical cavity provide an ideal solution for using reinforcements to suit the structural design requirements for Earth quake resistance (Figure 13 and Figure 14).

Figure 13. A Reinforced Masonry using interlocking blocks

Figure 14. The corner reinforced masonry with interlocking blocks

14. Boundary Walls

Dry stacked Interlocking block work is well suited for this application and is very fast, aesthetically pleasant and cost effective. Depending on height, width, area, application and other parameters, structural design can be made to adopt intermittent columns, coping band, foot stripping can be designed. Conduit blocks etc. for intermittent horizontal reinforcement to act as tie beam / lintel (Figure 15).

Figure 15. A boundary wall constructed with interlocking blocks

15. Advantages of Hydraform Interlocking Block Masonry

a) Mortarless blocks have high earthquake resistance due to interlocking and staggered position of units.

b) The joints being without mortar are flexible. The wall shall bear the shocks during the Earth-quake. It may be noted that the interlocking block technology is safer in earth-quake than the rigid brick/block masonry due to its flexible behavior. The energy generated during earthquake is dissipated due to interlocking non-rigid behavior.

c) Torsional strength of mortarless block wall is greater than that of brick wall or conventional block wall.

d) Due to greater percentage of air apace inside the blocks hollow spaces, it has a high insulation value.

e) The construction is very simple and normal workers can erect a wall with ease and speed.

f) There is a great flexibility of design because of the small size of basic units.

g) The block can be reused in any construction.

h) The blocks have sharp finished chamfered edges; therefore aesthetically the appearance of the wall looks pleasant. The external wall can be kept exposed finish with or without paint. Practically no plaster is required.

i) For the internal wall finish direct cement based putty can be used without cement plaster.

16. Limitations and Disadvantages of Hydraform Interlocking Block Masonry

a) The blocks shall be economical whether either the flyash (for flyash blocks) or particular type of soil (for soil blocks) is available nearby. However, blocks of cement-sand without flyash have also been used, but the blocks become heavy and sometimes uneconomical.

b) While chase cutting for conduits and water pipes etc., fine dust powder (due to fine flyash particles) is generated. Avoid inhaling the same.

c) It is essential to lay the corner blocks with cement slurry / thin cement-fine sand mortar for stability during erection of wall.

d) The wall should not be raised beyond lintel level unless lintel band is provided or the wall is loaded as the wall can topple due to high wind.

e) The gaps if any between the blocks may become source of water penetration.

f) In high rainfall areas, it may be necessary to finish external wall with cement plaster.

g) Hammering or drilling in wall may dislodge the blocks until the wall is loaded from above.

References

[1]   BMTPC, Techno-Economic Feasibility Report on Concrete Hollow and Solid Blocks, Building Materials and Technology Promotion Council (BMTPC), New Delhi, India, 2010.

[2]   P. K. Adlakha, New Building Materials, and Technologies, Vol. IV, Indian Building Congress, New Delhi, India, 2019.

Novel Approaches in Materials and Technologies for Building Construction   Materials Research Forum LLC
Materials Research Foundations **158** (2024)   https://doi.org/10.21741/9781644902936

CHAPTER 4

Reinforced Hollow Concrete Block Masonry

1. Introduction

The Reinforced Hollow Concrete Block Masonry (RHCBM) is an alternative cost-effective walling replacing Reinforced Cement Concrete (R.C.C.) framed structure. RHCBM can be used as load-bearing walls even for multi-storied buildings (upto 5 storeys) being constructed with hollow blocks. It is lightweight. With proper joints and reinforcement, the structure constructed with this wall can also be designed to be earthquake-resistant. With this load-bearing system, the traditional beams and columns can be eliminated. In this technology, the concrete blocks having two holes and block size 200mm x 200mm x 400mm are laid with staggered joints.

The vertical Reinforcement is placed through these holes as per design. Only those holes which carry the reinforcement are grouted with concrete, the others remain hollow (thus the building is lightweight). The concept of reinforced masonry utilizes the floors and roof as diaphragms acting as horizontal flanged girders to distribute lateral loads to walls. The walls provide horizontal shear resistance needed, in addition to carrying the normal vertical dead and live loads. This type of structure is defined as a Box system or Bearing wall System. For a building to be earthquake-resistant, the following three factors are to be taken into account:-

- The building should be lightweight.
- Structure behavior should be of box type load-bearing (i.e. the walls to act as shear walls and roof as the diaphragm).
- Openings and corners are stiffened with horizontal and vertical reinforcement.

Thus, the 200mm thick RHCBM wall acts as a shear wall with a slab as the diaphragm. As the blocks are load-bearing, the load is assumed to be evenly distributed on the footing. Continuity avoids any differential settlement also.

2. Reinforced Hollow Concrete Block Walls

The load-bearing structures whether of brick walls or hollow concrete block walls are safe, economical for normal dead and live loads as these loads are vertical and induce mostly compressive stresses. During earthquakes, cyclones, the horizontal (lateral) load becomes prominent and induces high tensile and shear stresses in the masonry walls. To make the structure earthquake and cyclone-resistant coupled with the economy, the walls are reinforced. By proper detailing of the reinforcement in the walls, it is possible to structurally integrate the walls to roofs and walls to walls and also ensure that there is shear wall action in every masonry element and the diaphragm action in the roof/floor

(Figure 1). The behavior of the structure is like a 3D-Box.Reinforcing a brick wall has some practical difficulties, that it requires some shuttering for reinforcing brick masonry. It breaks the continuity and integrity of masonry and that the reinforcement may get corroded.

Reinforcing hollow block masonry is much simpler and effective, because of the hollow spaces of the hollow blocks. The reinforcement can be provided vertically in the hollow spaces, without breaking the continuity of the masonry. The horizontal reinforcement is provided through the U-Channel block course. The reinforced hollow concrete block masonry (RHCBM) technique effectively utilizes the structural hollow concrete blocks for an efficient structural system, rather than its use only as infill in the RC framed structures. In principle, the reinforcement helps to develop the flexural tensile resilience and ductility needed for high-rise load-bearing structures [1].

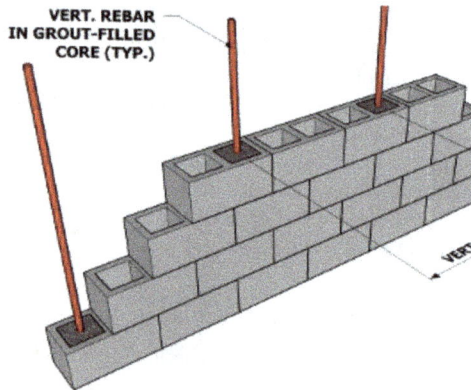

Figure 1. Graphical representation of reinforced hollow concrete block masonry

3. Properties of Mortar for Reinforced Concrete Masonry Walls

Mortar keeps concrete masonry units apart and keeps them together at the same time. Generally, the mortar thickness between units is about 9.5 mm but it could be modified to make rooms for concrete masonry dimension variations; in this way mortar keep units apart. Moreover, mortar holds concrete masonry units together by developing a bond between them through which flexural and direct tensile strength is produced. The tensile strength might not be significant in reinforced concrete masonry as in unreinforced masonry; nonetheless it is still crucial for the wall. A horizontal joint, which is called bed joint, is laid in face shell and mortar is placed on face shell only not the webs. Furthermore, vertical joints are called head joints, and mortar is poured into the head joints to a depth equal to the thickness of the face shell on both sides of the wall [2].

4.   Properties of Steel Reinforcement for Reinforced Concrete Masonry Walls

Steel reinforcement used for reinforced concrete masonry wall construction is same that is used in other reinforced concrete constructions. Joint steel bars as per ASTM A 951, is exclusive to masonry and it is galvanized to be protected against corrosion. The main task of joint reinforcement is to control shrinkage cracks. Steel bars are installed vertically in the concrete masonry cell and it is advised to employed spacer to place bars in the exact position. Reinforcement is placed in one layer so it installed at the center of the unit. Steel reinforcement can also be used on one side of the masonry units. For example, when lateral load like soil acts on the reinforced masonry wall or in the case of wind load if the wall is not subjected to wind suction. In either case, the designer should be aware to put steel bars in right position. For wall thickness of 30 cm or more, two layer of reinforcement can be used in seismic areas. Finally, reinforcement can be placed horizontally such as in walls that span horizontally between masonry columns embedded in the walls (Figure 2).

Figure 2. The construction of reinforced hollow concrete block masonry

5.   Structural Hollow Concrete Blocks

There is a difference between the hollow concrete blocks and the structural masonry blocks in terms of their dimensional consistency, quality, and strength. Normally, conventional hollow blocks are manufactured and used in place of bricks as infill in the RC framed system where bricks of good quality are not available; for a better quality of walling, fewer joints, and dimensional stability. But in such walls, their use is limited as infill walls only due to the lower strength of these blocks [3]. The structural hollow concrete blocks are of high strength and are used as load-bearing with adequate ductile detailing as shown in Figure 3.

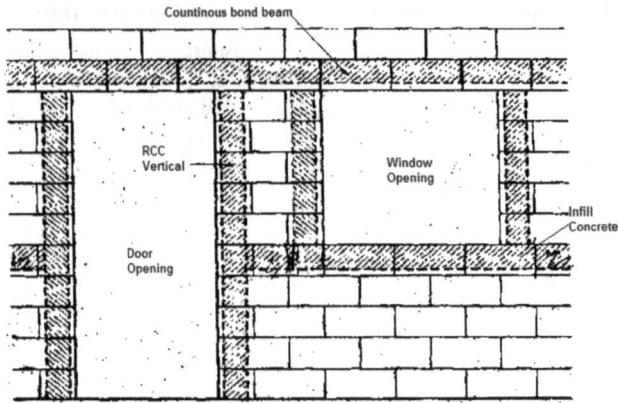

Figure 3. Structural bands for improved stability in RHCB

6. Structural Features of Reinforced Hollow Concrete Block Masonry

- The block is vertically reinforced with steel reinforcement and grouting concrete of minimum M-20 filled at regular intervals, through the common vertical cores of hollow concrete blocks.

- Each block is horizontally reinforced with steel reinforcement and grouting concrete-filled at plinth level, lintel level, and Roof level as, continuous R.C. band using U channel-shaped blocks in the masonry course.

- The reinforcement and grouting concrete is done without any shuttering and there is no scope for corrosion or reinforcement.

- As the reinforcement bars in both vertical and horizontal directions are continued into the roof slab and the walls respectively, the structural integrity is achieved in all three directions.

- In this technique, practically every wall and slab behave as a shear wall and a diaphragm respectively, thus increase in structural safety and stability.

- Due to the uniform distribution of reinforcement in both vertical and horizontal directions, increases the tensile resistance and ductile behavior of walls are increased, hence, the structure offers better resistance under dynamic loading.

- The technique is economical as it is a combined load-bearing and a shear wall structural system. Figure 4 shows a typical 3D cut-through view showing various features of RHCB Masonry.

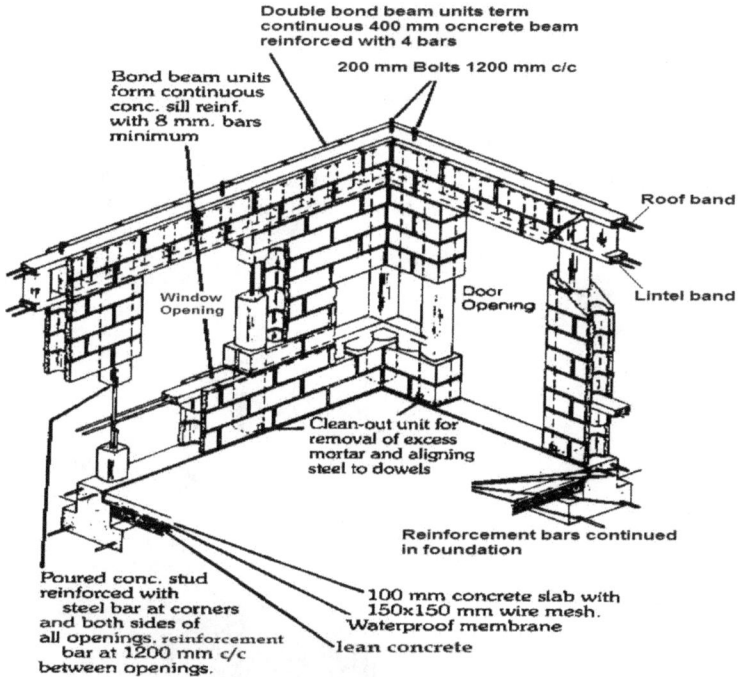

Figure 4. A 3D cut through view showing various features of RHCB masonry

7.   Precaution While Laying RHCB Masonry

The following precautions should be taken while laying RHCB Masonry:

a) The technology requires effective supervision and care during the tying of reinforcement, placement, grouting, and compaction.

b) The hollow concrete blocks should be of good quality concrete (Min. M-15)

c) The grout concrete should not be less than M-20

d) Care should be taken to place the vertical reinforcement in the center of the hollow space of the blocks.

e) For overlapping, tie the vertical bar first and then grout the hollow cores.

f) There is a practical difficulty in the placement of hollow blocks over the vertical bars.

g) Use 12mm and down aggregate, as the grout concrete has to fill the narrow hollow cores.

h) Use plasticizers for grout concrete, as mechanical vibration cannot be done.

i) For horizontal reinforcement use U-Shaped blocks.

j) Continue the vertical bars into the next floor after lapping. Provide extra bent-up bars and bend into the slab to structurally integrate the wall to the roof (Figure 5).

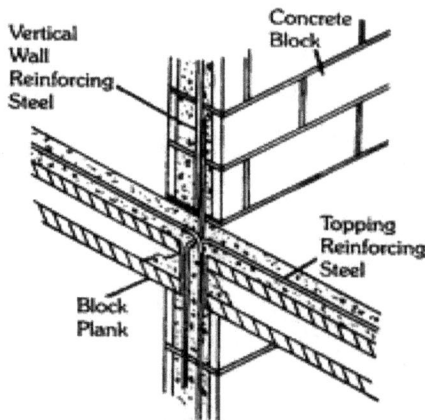

Figure 5. Wall and slab junction

k) There is a practical difficulty in the placement of hollow blocks over the vertical bars which require lifting of hollow blocks to the height of the vertical reinforcement end (Figure 6).

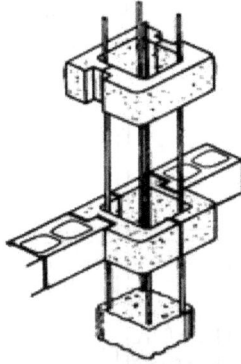

Figure 6. Placement of hollow blocks over vertical bars

l) For convenience, the blocks with split cut in one face can be manufactured and they can be placed at ease (Figure 7).

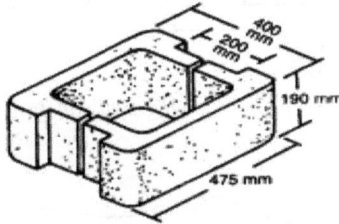

Figure 7. Split cut blocks

8.   Design Details of Reinforced Hollow Concrete Block Masonry

Figure 8 shows an example of two-bedroom twin flats marked as A and B in a 4 storeyed residential block with load-bearing reinforced hollow concrete block technology. Figure 9 shows the longitudinal section of the common wall with no openings (marked as W7), showing vertical reinforcement from below the plinth beam. Figure 10 shows the longitudinal section of wall W-16 with Door opening (marked in plan) showing vertical reinforcement for the Ground floor.

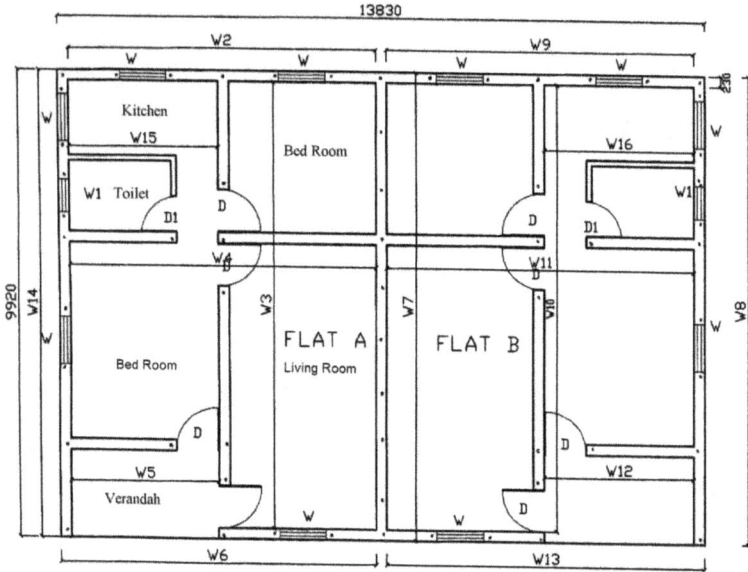

Figure 8. Typical Floor Plan showing wall panels for Design Analysis

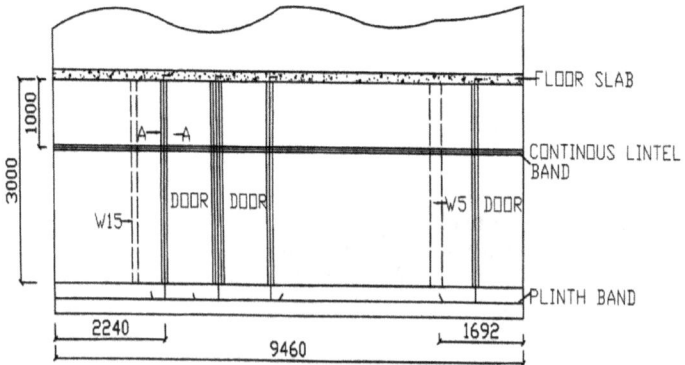

Figure 9. Sectional Elevation of Wall W3 at Ground Floor

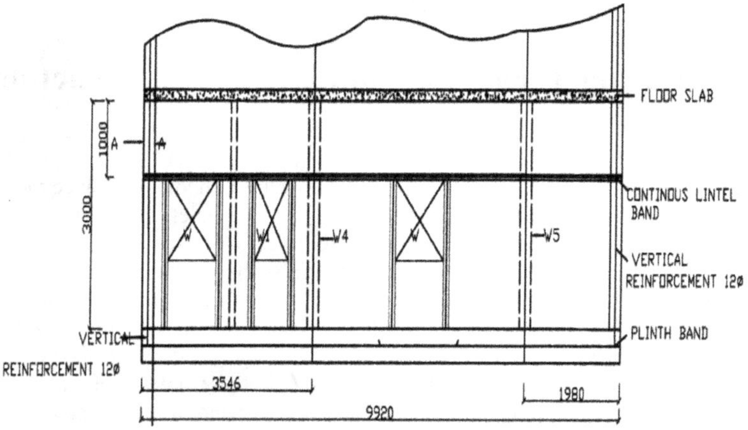

Figure 10. Sectional Elevation of Wall W14 at Ground Floor

References

[1]   P. K. Adlakha, New Building Materials and Technologies, Vol. IV, Indian Building Congress, New Delhi, India, 2019.

[2]   Jehkie, 2023, Information on https://blog.naver.com/PostView.nhn?blogId=jehk&logNo=221155450522

[3]   G. N. Kamat, Reinforced concrete block masonry (RCB) as an earthquake-resistant structure, 2013, Information on https://aceupdate.com/2013/10/17/green-buildings-by-innovative-masonry-method/

Novel Approaches in Materials and Technologies for Building Construction
Materials Research Foundations **158** (2024)

Materials Research Forum LLC
https://doi.org/10.21741/9781644902936

CHAPTER 5

3D Printing Technology in Building Construction

1. Introduction

3D printing is currently one of the world's fastest growing technologies. Since the 1980s, the notion of 3D printing has progressed, although few studies have concentrated on specific 3D printing technologies. Concrete's merits as a construction material are its durability and capacity to withstand natural calamities such as rain, snow, and wind while also providing shelter. Dr. Behrokh Khoshnevis, a researcher from the University of Southern California, developed a system called Contour Crafting (CC) in mid-2000s that paved the way for the present day's 3D Printung Technology and 3D Concrete Printing (3DCP). Various research institutes have commonly used Contour carving to create enormous 3D printed structures. The usage of contour carving in the construction sector can reduce the amount of physical labour required for projects while also lowering construction waste. Later in this study, the effectiveness of 3D printing concrete is discussed. The 3D printers are becoming increasingly popular as a result of these trends. In the realm of architecture, it has also caused a severe changes and transformation in various fields. Their application in architecture has evolved from scale modeling to a full-size final product [1]. One of the most significant changes in the world of architecture construction is the ability of using mock-ups as the final product [2]. Applications, benefits, methods and types of 3D printing, as well as their viability in employing concrete as a material in the process of manufacturing 3D printed concrete goods, have been examined in this chapter.

3D printing is a technique for creating actual objects by layering materials based on a digital model. All 3D printing methods necessitate the collaboration of software, hardware, and materials. Charles W. created the first 3D printer in 1983, and 3D printing has become one of the fastest developing technologies in recent decades. It was an extremely sophisticated and expensive technology in its early days. Over time, 3D printing became more prevalent in everyday life, and printers were increasingly used in industrial settings. Many people will refer to 3D printing as 'additive manufacturing,' especially when it is used in a factory setting, and many others will use both terms interchangeably. 3D printing technology can be used to make everything from prototypes and simple parts to highly complex final items like aeroplane parts, life-saving medical implants, automobiles, and even artificial organs employing layers of human cells as the technology advances [3]. The many types of 3D printers use different technologies to process various materials in different ways. Selective laser sintering (SLS) is a 3D printing technology in which a laser is used to fuse powder particles together to create an object. The materials utilized in SLS technique are typically strong and flexible. Nylon or polystyrene are the most common material [4].

2.   3D Printing and Additive Manufacturing

Additive manufacturing (AM) is a novel way for creating three-dimensional objects by layering materials like plastic, metal, concrete, sand, and other materials. Since the mid-1960s, additive manufacturing has been used, and there have been numerous improvements in terms of materials used in the process. The utilization of a computer, 3D modeling software (Computer Aided Design or CAD), machine equipment, and layering material are all prevalent AM methods [5]. After creating a CAD sketch, the file is uploaded to AM equipment, which reads the data from the CAD file and begins layering the material on top of each other to construct the thing in three dimensions. Powder, liquid, or sheet metal can all be utilized to make these objects. Rapid Prototyping (RP), Direct Digital Manufacturing (DDM), Layered Manufacturing (LM), and 3D printing are all examples of Additive Manufacturing methods that can quickly generate identical products. [5]. Additive manufacturing, often known as 3D printing, has proven to be very effective in lowering manufacturing costs and reducing waste during the manufacturing process. The fact that 3D printers are now widely used in a variety of industrial industries is due to the numerous advantages that this technology offers. The percentages of several disciplines, ranging from automobiles to medical, academic works to a variety of others are shown in Figure 1. Furthermore, according to various annual assessments, the market for 3D printing is predicted to increase fast in response to existing demands.

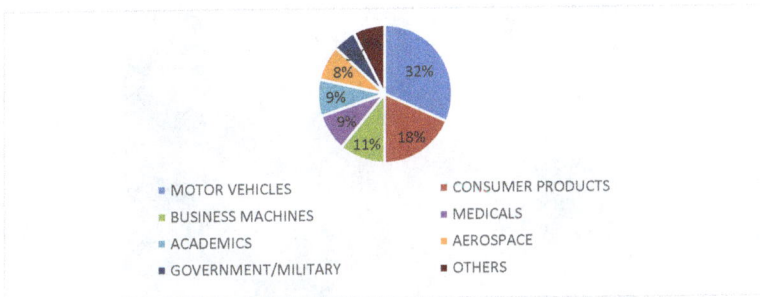

Figure 1. The range of 3D printing usage according to disciplines

Figure 2 shows the worldwide 3D printing industry forecast. If 3DP applications grow rapidly over the next ten years, concerns regarding the long-term viability of 3D printing manufacturing techniques may arise. As a result, research into the long-term viability of 3DP should be conducted before the markets burst, allowing for early adjustments [6].

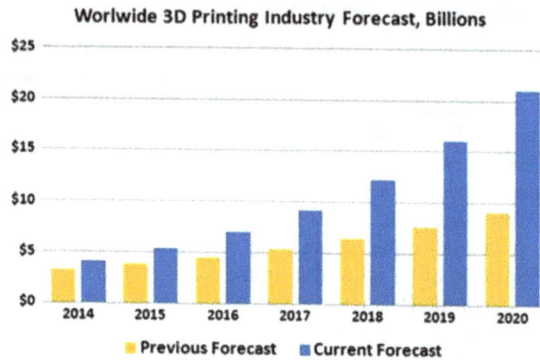

Figure 2. The worldwide 3D printing industry forecast

3D printing, in contrast to traditional manufacturing processes that are subtractive, is an additive method of production (Figure 3). In this regard, 3D printers may produce with a wide range of different material types, all of which are given in various stages (powder, filament, pellets, granules, resin etc.).

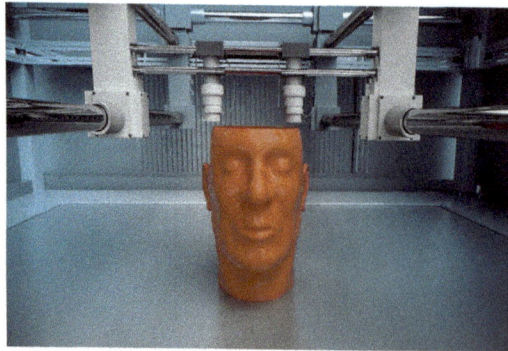

Figure 3. An example of Additive manufacturing

3. 3-D Printing Technology in Building Construction Industry

The comparison between traditional construction procedures and 3DP construction is shown in (Figure 4). Human resources are involved in many stages of traditional construction, which is time consuming and costly, as can be shown. In addition, the finished product contains a substantial amount of building waste. 3D printing, on the other hand, is largely automated and prints buildings from a computer-aided design (3D-CAD) with minimal human intervention and without the use of tooling, dies/formwork, or fittings [7]. As a result, many manual operations, labour needs, and material waste are reduced using this technology.

In traditional construction, the usage of formworks accounts for a significant amount of money spent, resulting in higher labour, equipment, and material expenses, as well as material waste [8]. Even for a simple geometrical arrangement, formwork contributes for 25% to 35% of the structural work cost, as well as a substantial amount of execution time [9]. Furthermore, the complexity of the geometrical configuration and type of formwork utilized have a detrimental impact on formwork labour productivity [10]. The productivity of 3DCP, on the other hand, is unaffected by the geometry of the construction architecture, as previously stated. There are many obstacles in the 3DCP procedures that researchers are still attempting to overcome in order to make 3D-printed buildings a better option than traditional construction (Figure 4). The printing systems, material qualities, and competent monitoring of the process are all critical to the success of 3DCP.

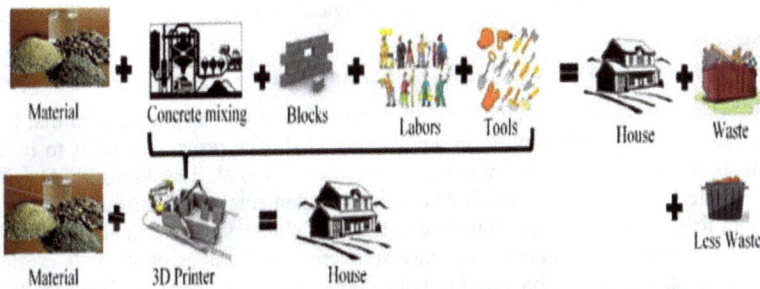

Figure 4. Conventional construction vs 3D printing construction process

The printing systems are primarily automated technologies that use huge 3D printers to manufacture large-scale construction elements. According to a Tractica, a rising number of construction organizations are using robots to address the skilled labour shortage and reap the benefits of improved speed, efficiency, safety, and profits. According to this increased interest in construction robots, more than 7000 construction robots will be deployed between 2018 and 2025, with market revenue rising from USD 22.7 million in 2018 to USD 226.0 million in 2025 [11].

4.   The 3-D Printing Technology in Architecture and Construction

3D printing architectural models has emerged as a viable option. Architectural models are typically formed of cardboard, wood, or other easily moldable materials. Architects require models in order to research many parts of their designs. In both architectural and interior design, it is frequently changed to get a flawless concept of their vision. 3D printing on a large scale in the building industry will have a wide range of uses in the private, commercial, industrial, and public sectors. Faster construction, cheaper labour costs, improved complexity and/or accuracy, greater function integration, and less waste are all potential benefits of these technologies. Several different approaches have been demonstrated to date, including on-site and off-site manufacture of buildings and construction components, as well as the use of industrial robots, gantry systems, and tethered autonomous vehicles [12]. Fabrication of dwellings, construction components (cladding and structural panels and columns), bridges and civil infrastructure, artificial reefs, follies, and sculptures have all been demonstrated utilizing construction 3D printing technologies. It may also allow construction in tough locations where a human labour is not acceptable, for as in places with specific circumstances [13].

5. Case Studies

The various modern trends in 3D printing technology with reference to building construction system that have been employed in contemporary architecture around the world through three case studies have been analyzed. The following three case studies discuss and illustrate how various technological concepts have been realized architecturally.

5.1 Case Study 1: 3D Canal House, Netherlands

The 3D Print Canal House is a DUS Architects research project in Amsterdam that looks into the possibilities of 3D printing in architecture [14]. The projects' goal is to build a 13-room show house. In a shipping container beside the Canal, a customized 3D printer named "KamerMaker" was installed. Many printed elements make up the house. Each component features a new research update in terms of shape, structure, and material. The project demonstrates how architecture may spur cross-sectoral innovation. It is jointly sponsored by all partners, who provide both knowledge and financial resources to the project [15]. (Figure 5).

Figure 5. General view of the project

Materials Research Forum LLC
https://doi.org/10.21741/9781644902936

The printer uses a bio-plastic blend of plastic fibres and 80 percent plant oil to build wall components. To give structural strength, the wall components are linked together and filled with bio-concrete. It's vital to highlight that the entire project's materials are recyclable (Figure 6, 7 and 8). The canal hose has the potential to change the building business and offer new tailor-made dwelling alternatives around the world as an entrepreneurial building project [15, 16].

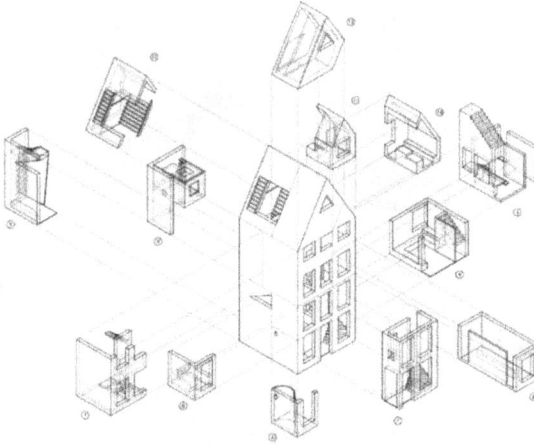

Figure 6. Decomposition of the model

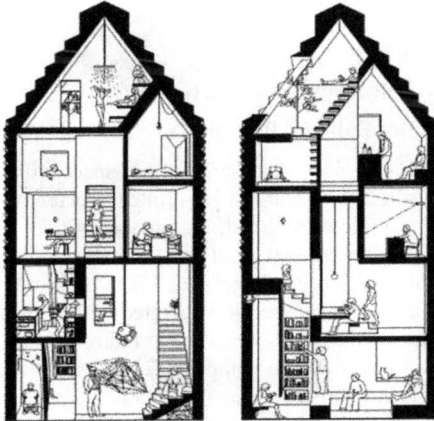

Figure 7. Section of 3D Canal House, Netherlands

Materials Research Forum LLC
https://doi.org/10.21741/9781644902936

Figure 8. 3D printed building blocks

This effort grew out of those "research and do" concepts, according to the interdisciplinary research team:

- KamerMaker: large scale 3D printing
- Sustainable 3D print material for the building industries
- New construction and building techniques
- Downloadable tailor-made architecture
- Smart building
- Scripted city planning

In 2017, the project was finally constructed and completed [17].

5.2   Case Study 2: Office Building By Apis Cor, Dubai, UAE

Apis Cor's 9.5-meter-high and 640-square-meter-floor-space office building (Figure 9 and Figure 10) in Dubai is the world's largest 3D-printed structure to date [18]. It is stated that the construction was completed with half the number of craft workers and produced 60% less waste [19]. The building was built far faster than a traditional construction project, especially the on-site 3D-printed sections, which took only two weeks [20, 21]. The printing time for the general contractor integrated ceiling, electronics, and fittings was roughly two months. Furthermore, when the printing process was completed, the project required an extra two months to complete [22]. Table 1 shows several building components and their construction processes in general. As can be observed, all of the walls that account for the building's principal concrete works were 3D printed. Furthermore, column formworks were printed, which saved a great deal of time and formwork materials. The remainder of the construction work was completed using traditional methods. As a result, the potential savings from 3D printing, such as a 50-70 percent reduction in construction costs, a 50-80 percent reduction in labour costs, and a 60 percent reduction in material waste [23], are only applicable to the 3D printed building components, not to the entire construction project.

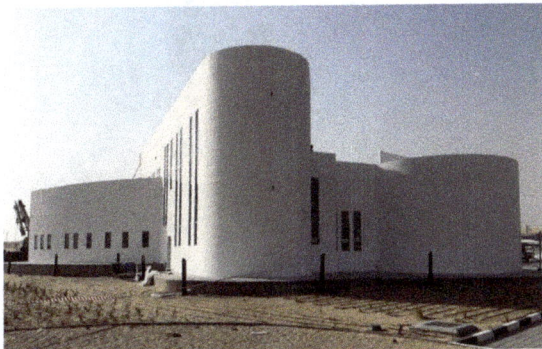

Figure 9. World's largest office building in Dubai by Apis Cor

Figure 10. Construction of the office building in Dubai

Table 1. Construction method of various components of 3D printed building in Dubai

Building Components	Construction Method
Foundation Walls	Conventional construction method by general contractor 3D printed
Columns	Reinforcements were placed manually Formworks were 3D printed Concreting was done manually
Slab Roof, ceiling, windows, insulations Plumbing and electrical installation	Precast Conventional method by general contractor Conventional method

5.3   Case Study 3: TECLA By Mario Cucinella Architects and Wasp, Italy

Mario Cucinella Architects in Bologna has teamed with 3D printing specialists WASP to build a low-carbon home prototype that is 3D printed using clay (Figure 11).

Figure 11. Exterior view of the TECLA house

The residence was designed by Mario Cucinella Architects and built and developed by Wasp using clay collected from a nearby riverbed. It was named Tecla, a mix of the terms technology and clay. An Italian architecture team was inspired by the potter wasp and utilized 3D printing to create the domed, beehive-like structure of a house as depicted in plan and section, (Figure 12 and 13) out of zero-emissions clay in the hopes of demonstrating the technology's potential for sustainability [24].

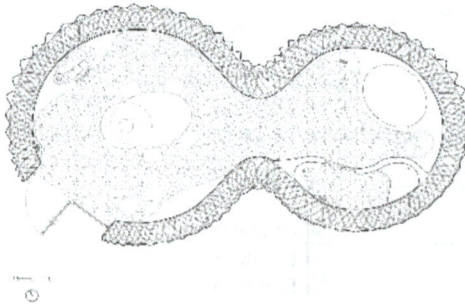

Figure 12. Plan of TECLA house

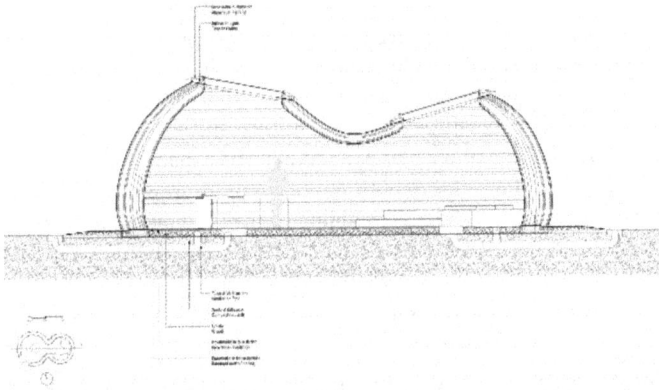

Figure 13. Sectional details of TECLA house

The buildings are created with clay from wherever they are being built, just like the industrious wasps, which means that if they have to be demolished, the only waste is the plumbing, gas, and electrical components. Cucinella co-developed the approach, TECLA, which stands for technology and clay, with the support of another firm called WASP, which specializes in 3D-printing solutions. Their modular design uses two 3D-printing arms at the same time to generate two domed rooms out of 350 layers of undulating clay with rice chaff as insulation, comparable to the traditional Moroccan Kasbah building methods. The goal is to be completely self-sufficient, and the design and durability may be adjusted to meet local climate and environmental challenges. Cucinella missed out on calling the project "circadian cupulas" since one is meant for the day, with a wide circular skylight and door letting in lots of natural light, while the other is designed for the night, with a smaller, warmer, enclosed setting under a smaller window [25]. According to Treehugger, Cucinella Architects plans to build these cupulas as self-sustaining eco-communities for both the fringes of cities and undeveloped countries, thanks to WASP's contribution of a DIY-version of the dwellings.

6. Advantages and Disadvantages of 3D Printing in Building Construction

Curvilinear forms, rather than rectilinear forms, can be used to arrange structures in 3D printed buildings, making them far more robust. Walls of structures are typically hollowed out to allow utility lines to pass through while also reducing the quantity of materials utilized. Less material not only adds to the durability of the product, but it also saves architects and builders money on building costs. While the benefits are limitless, there are still a number of roadblocks in the way of this technology's adoption. Let's look at the benefits and drawbacks of 3D printing construction [26].

6.1   Advantages of 3D Printing in Building Construction

a) **Construction waste is reduced:** It is more environmentally favorable to use 3D printed construction for architectural projects. The printed construction will only generate roughly 30% of the garbage that a typical construction project produces while utilizing very little energy.

b) **Increased the Number of Design Shapes:** 3D printing allows for design shapes and customizations that would be impossible or prohibitively expensive to achieve through traditional building. Small amounts of concrete may be precisely placed where needed for complex shapes by printers, considerably improving an architect's design choices.

c) **Reduced Construction Time:** When a building project is completed with a 3D printer, the construction time is drastically reduced.

d) **Construction Costs are Reduced**: Because of the savings in raw materials and, more crucially, labour, 3D printed buildings are much less expensive to construct than those constructed using traditional methods. By having most of the building performed by 3D printers, an architectural project's labour expenses can be lowered by up to 80%.

6.2   Disadvantages of 3D Printing in Building Construction

a) **Building Regulations:** There are no rules or procedures for obtaining approval for 3D printed structures for home or commercial usage. First, the government would have to establish electrical, plumbing, structural integrity, and public safety rules that must be obeyed [26].

b) **Types of Materials:** Concrete and polymers are about the only materials that can be delivered from the printing head.

c) **Engineering Compatibility:** 3D printed buildings have piqued the interest of a small number of architects and engineers. During the design process, the additional capabilities that come with the new technology are not utilized. Because traditional plans aren't compatible with 3D printers, the entire design process must be rethought.

d) **3D Printing Technology is Expensive:** The cost of 3D printing equipment and materials makes the technology prohibitively expensive. Industrial 3D printers are still quite expensive, costing hundreds of thousands of dollars, making the technology's initial costs very high. Capital investment for a single machine starts in the tens of thousands of dollars and can reach hundreds of thousands of dollars.

e) **3D Printers aren't that User-friendly:** 3D printers have come across as easy to operate and sound more useful than they really are due to the excitement and possibilities surrounding 3D printing technology.

Materials Research Forum LLC
https://doi.org/10.21741/9781644902936

f) **Unemployment** Because 3D printing just requires one process, it may produce product ideas and prototypes in a couple of hours. It gets rid of a lot of the phases that subtractive manufacturing requires. As a result, it does not necessitate a significant amount of labour. As a result, 3D printing may result in fewer manufacturing jobs. Manufacturing employment losses might have a significant impact on economies in countries that rely heavily on low-wage jobs.

7.   Future of 3D Printing in Construction

Although 3D printing technology (Stereo Lithography, Fusion Deposition Modeling) has been around since the early 1980s, it has been reluctant to catch on. There are many different perspectives on where the technology might go, but it will become more cost efficient in a variety of industries, making research and mass customization in architecture more possible. Although 3D printing is still in its infancy across all industries, the technology's prospective benefits appear to be propelling it forward. Some predict that 3D printers would mostly be used to manufacture building components and panels in factories or on-site, while others see 3D printing as a transformative technology that will alter the construction sector.

7.1   3D Concrete Wall Printing Technology in India

There are various disadvantages to India's conventional construction method. The long time frame, big number of utilized labour, high percentage of error, and long-distance transportation all demonstrate this. All of these difficulties have a negative influence on the environment and raise development costs. The emerging new technique of three-dimensional printing can effectively overcome such drawbacks.

7.2   Possibilities and Challenges

According to discussions, construction time efficiency may be a crucial role in the introduction of this technology in India, but the type and scale of the structure are also important factors to consider. As a result, it is possible to argue that 3DP technology will be successful in both mass manufacturing projects and massive buildings with restricted floors. Has the potential for speedy production and high quality, which could compensate for the current low quality of work among working people. Furthermore, the adoption of this technology will allow certified architects to reclaim control, resulting in improved internal environmental quality and energy performance in the construction industry across the country. Furthermore, modifying India's current building standards may increase architectural identity and strengthen architects' final responsibilities in the construction business. When comparing the cost of bricks (used for ordinary wall construction) to the cost of reinforced concrete (which should be utilized for 3DP), the projected hurdles to the use of this technology is linked with cost [27]. More obstacles include a lack of practical knowledge, a lack of scientific information, and industry/society rejection of norm-changing approaches, as well as a lack of stakeholder interest, a skilled workforce, and the cost of importing. Another factor to consider is modification restrictions, as working with established models limits design flexibility. It's also unable to receive change orders during construction, thus design changes are out of the question. Modifications are also not possible during operation. Other factors to consider include the high costs of durability and maintenance operations. Furthermore, the primary source of worry is that 3DP technology is imported rather than developed locally, resulting in significant cost increases.

Conclusions

With rising urbanization in many developing countries, creative ideas to improve the long-term performance of the structures in which people live and work are urgently needed. The rapid advancement of 3D printing technology has changed people's minds about using concrete as a 3D printable material on its own. 3D printing is a disruptive technology that can be used in the construction industry to gain economic and environmental benefits. It is an automated layer-by-layer production method. In a wide number of architectural research topics, 3D printing technology offers various economic benefits. To begin with, 3D printers can theoretically build most objects that can be redrafted using CAD software rapidly and cheaply. This has a significant practical benefit for daylighting studies as well as many other sorts of empirical architecture studies. Second, compared to traditional manufacturing, it results in less waste of raw materials. Third, unlike traditional manufacturing, which relies on large-scale production to offset costs, 3D printing is very cost-effective for smaller production runs or large-scale customization, including the creation of extremely intricate, difficult, or impossible-to-make products using traditional manufacturing techniques. The industry is pursuing a number of initiatives aimed at lowering the cost of technology.

This study has reviewed the many commercially available 3D printing systems in terms of their basic benefits and limitations, as well as the different types of 3D printing concrete technologies, as well as research and commercial projects that have used the technologies. The numerous options for employing concrete as a 3D printing material, as well as the technology's future, have been examined. The process of large-scale 3D printing of cementitious materials is examined. The advantages and disadvantages of using concrete in 3D printing have been examined. The findings of this article, particularly the analysis of case studies, show that 3D printing technology has significant architectural potential. These innovations have the potential to transform the construction industry's future. It is feasible to assert that if development continues at a given rate, it will change the construction process. However, in order for technology to advance at a faster rate, more resources are required. The truth is that none of the 3D construction printing projects that have been completed so far have been competitive. There may have been some cost savings in labour and materials, but it was not competitive with standard construction methods.

References

[1]   S. A. Selçuk, A. G. Sorguç, Reconsidering the role of biomimesis in architecture:

an holistic approach for sustainability, 2nd International sustainable building symposium, Ankara, 2015, pp. 382–388.

[2]   A. Selçuk, G. Sorguç, B. E. Şantiyeye, Y. Dergisi, Turkey, 2015, pp. 154–160.

[3]   I. Hager, A. Golonka, R. Putanowicz, 3D printing of buildings and building components as the future of sustainable construction?, International Conference on Ecology and new Building materials and products, Cerna Hora, 2016.

[4]   3D Printing Industry, The free beginners guide [Online] Accessed on: 13 December 2021 Information on https://3dprintingindustry.com/3d-printing-basics-free-beginners-guide/technology/

[5]   S. Chang, H. Shawn, Exploring the evolution of additive manufacturing industry: a study of stakeholder requirements and architectural analysis of Desktop three-dimensional printing Segment, Massachusetts Institute of Technology, Massachusetts 2016 Information on https://dspace.mit.edu/handle/1721.1/106243

[6]   M. Gebler, A. J. S. Uiterkamp, C. Visser, A global sustainability perspective on 3D printing technologies, Energy Policy, Elsevier, 2014, 74, pp. 158–167.

[7]   Y.W.D. Tay, B. Panda, S. C. Paul, N. Mohamed, M. J. Tan, K. F. Leong, 3D printing trends in building and construction industry: A review, Virtual Phys. Prototype, 2017, 12, pp. 261–276.

[8]   D. Camacho, P. Clayton, W. J. O'Brien, C. Seepersad, M. Juenger, R. Ferron, S. Salamone, Applications of additive manufacturing in the construction industry-A forward-looking review" Autom. Constr., Elsevier, 2018, 89, pp. 110–119.

[9]   V. Mechtcherine, V. N. Nerella, F. Will, M. Näther, J. Otto, M. Krause, Large-scale digital concrete construction-CONPrint3D concept for on-site, monolithic 3D-printing," Autom. Constr., Elsevier, 2019, pp. 107.

[10]   A. M. Jarkas, Effect of Buildability on Labor Productivity: A Practical Quantification Approach, Journal of Construction Engineering Management. 2016, 142, 06015002.

[11]   Tractica, Construction Robotics Market to Reach $226 Million Worldwide by 2025, 2019, Information on https://tractica.omdia.com/newsroom/press-releases/construction-robotics-market-to-reach-226-million-worldwide-by-2025/

[12]   S. C. Paul, A review of 3D concrete printing systems and materials properties: current status and future research prospects, Rapid Prototyping Journal, 2018.

[13]   N. Nithesh, Development of concrete 3D Printing, Aalto University, Finland 2018.

[14]   E. Chalcraft, Amsterdam Architects Plan 3D – Printed Canal House, Dezeen, 2013, Available: https://www.dezeen.com/2013/03/09/amsterdam-architects-plan-3d-printed-house/

[15]   DUS Architects 3D print canal house, New Material Awards, 2014, Available:

https://new-material-award.nl/en/3d-print-canal-house/

[16] A. C. Van der Veen, The structural feasibility of 3D-printing houses using printable polymers, Doctoral [dissertation], TU Delft, Delft University of Technology, Netherlands, 2014.

[17] V. Quirk, First 3D Printed House to Be Built in Amsterdam, ArchDaily, 2021, Information on https://www.archdaily.com/491666/first-3d-printed-house-to-be-built-in-amsterdam

[18] Apis Cor, The world's largest 3D printed building, Apis Cor, 2020, Information on https://www.apis-cor.com/dubai-project

[19] V. Carlota, Apis Cor Collaborates on World's largest 3D Printed Building, 3D, 2019, Information on https://www.3dnatives.com/en/apis-cor-largest-3d-printed-building-261020194/#!

[20] N. Jewell, World's Largest 3D-Printed Building Opens in Dubai after 2 Weeks of Construction," Inhabitat, Information on Available: https://inhabitat.com/worlds-largest-3d-printed-building-opens-in-dubai-after-2-weeksof-construction/

[21] E. Bravo, An Administrative Building in Dubai, the largest 3D Printed Structure in the World, Available: https://www.smartcitylab.com/blog/digital-transformation/largest-3d-printed-building-in-the-world/

[22] M. Molitch-Hou, World's largest 3D Printed Building Unveiled in Dubai, 3D prints, 06 January 2020 [Online] Accessed on: 08 Dec 2021. Available online: https://3dprint.com/261978/worlds-largest-3d-printed-building-unveiled-in-dubai-d/

[23] Arch 20, 25% of Dubai Buildings Will be 3D Printed by 2030 Information on https://www.arch2o.com/25-dubaibuildings-will-3d-printed-2030/

[24] J. Parkes, Tecla house 3D – Printed from locally sourced clay, Dezeen, 2021. Information on https://www.dezeen.com/2021/04/23/mario-cucinella-architects-wasp-3d-printed-housing/

[25] A. Corbley, World's first 3D Printed House made of local raw Earth – And it closes the roof with a dome, Goodnewsnetwork, 2021, Information on https://www.goodnewsnetwork.org/tecla-sustainable-3d-printed-houses-from-cucinella-architects/

[26] R. J. M. Wolfs, F. P. Bos, and T. A. M. Salet, Early age mechanical behavior of 3D printed concrete: Numerical modeling and experimental testing, Cement and Concrete Research, Elsevier, 2018, pp. 103–116.

[27] S. Ahmed, 3D Printing in Architecture, Engineering and Construction, Engineering Research Journal, 2019.

Novel Approaches in Materials and Technologies for Building Construction   Materials Research Forum LLC
Materials Research Foundations **158** (2024)   https://doi.org/10.21741/9781644902936

CHAPTER 6

Structural Insulated Panels

1. Introduction

Structural Insulated Panels (SIPs) are high-performance building systems for residential and light commercial construction. The panels consist of an insulating foam core sandwiched between two structural facings of rigid board sheathing materials, typically Oriented Strand Board (OSB). The foam core is generally one of the following: Expanded Polystyrene (EPS), Extruded Polystyrene (XPS), and Polyurethane Foam (PUR). With EPS and XPS foam, the assembly is pressure laminated together. With PUR and PIR, the liquid foam is injected and cured under high pressure. SIPs are manufactured under factory-controlled conditions and can be fabricated to fit nearly any building design. The result is a building system that is extremely strong, energy-efficient and cost-effective. The most common sheathing boards are Oriented Strand Boards (OSB). Other sheathing materials include sheet metal, plywood, fiber-cement siding, magnesium-oxide board, fiberglass mat gypsum sheathing, and composite structural siding panels [1]. SIPs take the form of an insulating core or layer sandwiched between two structural facings, as shown in Figure 1. Usually, this insulating core can be closed-cell polyurethane foam or expanded polystyrene (EPS). Oriented strand boards (OSB) are the commonly used structural facings as they comply with the British Standard BS EN 300 for structural integrity (Figure 2). Other than OSB, cement, metals, engineered woods, magnesium oxide board, and cement are also used as structural materials. SIPs are manufactured in various dimensions and thicknesses, ranging between 4 and 8 inches (Figure 3). They are manufactured in a factory and can be installed quickly as soon as they are delivered on-site. Based on the specifications of the project, SIPs can be customized and manufactured. For walls, SIP is manufactured with a maximum length of up to 7.5m, and for floors, it can reach up to 4m [2].

Figure 1. Typical SIP with oriented strand boards (OSB) and EPS

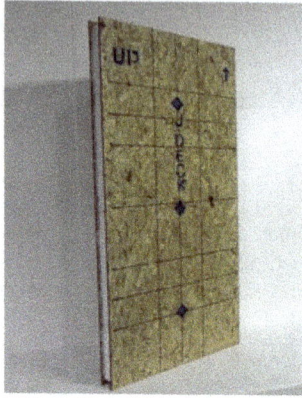

Figure 2. A full-size structural insulated panel

Figure 3. Sheet metal structural insulated panel

2.   Types of Structural Insulated Panels

Based on the insulating material used for manufacturing [2], there are mainly three types of SIPs:

2.1   Expanded Polystyrene S Structural insulated panels

Expanded polystyrene (EPS) is a foam board insulation widely used in the manufacturing of SIPs. It is a type of closed-cell insulation manufactured by expanding a small bead of polystyrene polymer using steam. These beads, during the process, convert into the form of large insulation blocks of varying densities that are later cut into the required shapes and sizes. These SIPs offer an R-value of R-4 to R-5 per inch of thickness. This value can go as high as 13.8. EPS panels are available in the market ranging between 4 to 24 feet in width.

2.2 Polyurethane Insulated Panels

Polyurethane or polyisocyanurate-insulated panels provide a nominal R-value of around R-6 to R-7 per inch of thickness. These insulated panels are manufactured at a thickness of 3.5 inches for walls and 7.5 inches for ceilings. Polyurethane panels are expensive compared to EPS but offer higher water and fire resistance and higher R-value.

2.3 Compressed Straw-Core Insulated Panels

Compressed straw-core insulated panels are "green building materials" manufactured using renewable and recycled waste agricultural straw, unlike other panel materials. Even though they are environmentally friendly, they offer a small range of R-value compared to other types of SIPs. The sheathing materials and foam types have their benefits and drawbacks. The type of SIPs selected depends upon the building type and site conditions. The following Table 1, Table 2 and Table 3 outline the benefits and drawbacks of the most common sheathing and foam types.

Table 1. Core Type Chart

Foam Core	Benefits	Drawbacks
Expanded Polystyrene (EPS)	Least expensive; thickness options are only limited by the foam manufacturer; availability; fastest to modify in the field; most benign blowing agent	Produced with HBCD*
Extruded Polystyrene (XPS)	Strength; water resistant	Availability; produced with HBCD*
Polyurethane Foam (PUR)	Highest R-value/inch; strength, water resistant	Most expensive; harder to modify thickness limitations; creep; availability; produced with chlorinated phosphate flame retardants

*HBCD: hexabromocyclododecane - a brominated fire retardant classified by the European Union (REACH program) as persistent, bioaccumulative, and toxic (PBT).

Table 2. Sheathing Type Chart

Sheathing Type	Benefits	Drawbacks
Oriented Strand Board (OSB)	Load bearing; readily available; tested; large panel size up to 8' x 24'	Subject to mold; reduction in structural capacity, if exposed to moisture; not fire resistant; must be treated for termites; difficult substrate for common joint tapes
Sheet Metal	Resistant to mold; can be load-bearing; very light; unlimited lengths when made from coil stock	Must be galvanized or stainless steel; not load bearing
Plywood	Lateral strength	Availability; price; limited panel size; subject to mold and reduced structural capacity, not fire resistant; must be treated for termites
Fiber Cement Siding	Resistant to mold, termites, * fire	Availability; weight; testing; limited panel size
Magnesium Board	Resistant to mold, termites, & fire	Availability; testing; limited panel size
Fiberglass Mat Gypsum Sheathing	Resistant to termites & fire	Not structural; limited panel size
Composite Structural Siding Panels	Resistant to mold and termites; pre-primed materials available	Not fire resistant

Table 3. Technical data of different foam

Foam Type*	EPS Foam	XPS Foam	PUR Foam
Density in Panel (lb/ft^3)	0.90	1.5	2.3 – 2.5
Compressive Strength @10% deformation(psi)	10	20	35
R-value/in @ 75° F	3.6	5.0	6.54
Permeance per inch	5	1.1	2.00
Common Fire Retardant	HBCD	HBCD	TCPP
Common Fire Rating Class	1	1	1
Common Blowing Agent	Pentane	HFC—134a	HFC—245fa

* Most SIP manufacturers use a 0.95 minimum density.

3. Structural Design and Construction with Structural Insulated Panels

SIPs behave similarly to a wide flange steel column in that the foam core acts as the web and the sheathing responds as the flanges. Under axial loads, the sheathing responds similarly to a slender column, and the foam core acts as continuous bracing preventing the panels from buckling. Just as wide flange sections increase in strength with increased depth; thicker cores result in stronger panels in compression and bending [3]. SIPs are designed to resist not only axial loads but also shear loads and out-of-plane flexural loads. The panels' ability to resist bi-axial bending and lateral shear allows them to be used as roofs and floors. SIPs panels are acceptable to use as shear walls in all seismic design categories. A structural engineer should determine if a secondary structural system is required based on the design loads. To date, the tallest structure constructed exclusively of SIPs is four stories. Taller structures are possible; however, design limitations are since SIPs are bearing walls and therefore open spaces on lower floors are more difficult to achieve (Figure 4). Often large SIPs structures rely on a secondary framing system of steel or timber to satisfy requirements for unobstructed spaces. Unique screw connections are available to attach SIPs to wood, light gauge steel, and structural steel up to 1/4 inch thick. Foundations for SIPs panels must be level. There is little tolerance for differential settlement. If there is a substructure shift, it will compromise the sealant of the panels' joints which may cause moisture infiltration. Allowable deflection tolerances set by the manufacturer of the panels and sealants should be consulted when designing the foundation. Minor imperfections may be accommodated with careful, skilled installation.

Figure 4. SIP ridge and gable end tape/gasket seal

Joint design is imperative for structural and long-term durable performance. One particular weakness of SIPs panels is air penetration from the interior at joints or penetrations. In cold climates, if warm humid interior air reaches the interior face of the outer sheathing layer it can condense, causing rot and deterioration. Frequently this outer layer is Oriented Strand Board (OSB), which is particularly susceptible to moisture damage. Proper joint design should be given special attention, and if properly executed in the field, will eliminate the air infiltration problems. The primary joint design generally includes seals within the thickness of the panel, typically spray foam or gaskets. There should be an overflow seeping of the spray foam at the joints to indicate a full-depth joint seal as shown in the Figure 5. An additional secondary seal air seal of tape or gasket should be provided at the interior face of the panel, especially in cold climates.

Figure 5. SIPs used for roof panels showing seeping of sealant at SIP joint

Figure 6. SIPs used for wall and roof panels showing seeping of sealant at SIP joints

Two of the most widely used panel joint connections are the surface spline and the block spline. The surface spline joint connection consists of strips of OSB or plywood inserted in slots in the foam just inside each skin of the SIP. The block spline is a thin and narrow SIP assembly that is inserted into recesses in the foam along the panel edges. The surface spline connection and the block spline connection result in a continuous foam core across the panels, eliminating air infiltration at the joints. If structurally required, panel joints can be reinforced with one or more 2x lumber studs or Laminated Veneer Lumber (LVL) along the edges of the two panels to be connected. One disadvantage of this type of connection is that a thermal bridge is created at the joint (Figure 6). Another joint connection, mechanical Cam locks, create a tighter joint between panels but makeup only a small percentage of the market. In addition, Cam locks can only be set in PUR because the locks require a higher tensile strength than provided by other foams and the foam needs to expand and set around the lock's flanges. In any type of connection, the seam along the sheathing must be covered with a continuous line of foam sealant and/or panel tape.

Openings can occur anywhere within the panel, including at the edges and corners. The panel foam can be recessed to accept 2x lumber headers. However, panels can be reinforced at headers so that additional structure is not required during construction. The inside panel and foam can be subtracted to provide beam pockets for roof and floor joists. Any opening within the SIPs that accepts another enclosure element must be properly sealed. Plumbing chases are usually located in furred-out framing or conventional framing should be used for plumbing walls. Electrical chases 1-1.5 inches in diameter can be incorporated into the SIPs during the manufacturing stage. Foam is applied in any gaps resulting after the installation of electrical wiring (Figure 7).

Figure 7. SIPs used for a furred-out wall at an exterior panel for vents and plumbing

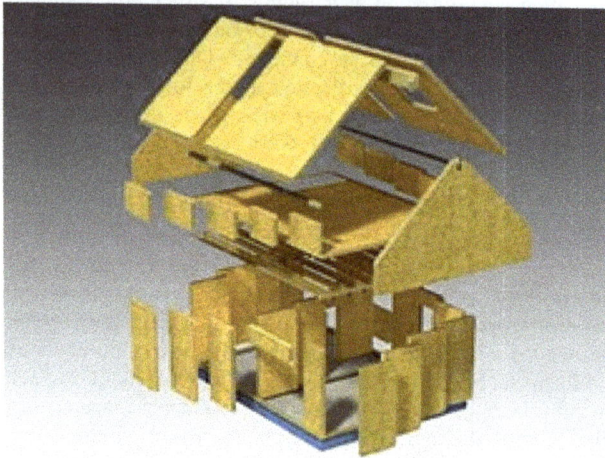

Figure 8. Exploded axonometric of a SIPs building

Other unexpected penetrations made in the panels during construction should be made 1 inch larger in diameter than the penetrating pipe to allow for the application of foam sealant. The typical wall panel thicknesses are 4.5 inches and 6.5 inches. The largest panel size to date is 9' x 24'. Curved panels are possible although not common and it is often more practical to use stud framing for non-orthogonal geometries (Figure 8). The roof panels are typically 10.25 inches and 12.25 inches thick. Roof panel thickness depends upon the required R-value and span. EPS and XPS panels can be made up to 12.25 inches thick. PUR and PIR panels can be made up to 8.25 inches thick. End wall panels for various roof profiles can be achieved with SIPs.

4.   Performance Issues

4.1   Thermal *Performance of* Structural Insulated Panels

The quality of a building's envelope is measured by its ability to prevent infiltration of outside air. Recent energy code standards require an air-tight building envelope, and a SIP building with properly sealed panel joints is inherently airtight. The results of blower door tests on a room with SIPs walls and ceilings, one window, one door, and pre-routed wiring chases and electrical outlets compared to an identical room of 2x6 studs, OSB sheathing, fiberglass insulation, and drywall showed the SIPs structure to leak 90 percent less than the stud structure.

The Whole Wall R-Value of a wall assembly is currently the most accurate method of quantifying its thermal performance (Table 4). The Whole Wall R-Value takes into account the resistance of heat flow through an opaque cross-sectional area of the insulation and structure while accounting for the loss of energy at the interfaces of the

Materials Research Forum LLC
https://doi.org/10.21741/9781644902936

wall with the roof and floor and at corners and fenestrations. The Whole Wall R-value of a 4-inch SIP wall is 14. The Whole Wall R-value of a 2x4 wall is less than 10. The Whole Wall R-value of a 2x6 wall is between 11 and 13.7 depending on the quality of the installation of batt insulation. The elimination of thermal bridging and a more air-tight envelope contribute to the higher Whole Wall R-value of SIPs walls when compared to conventional metal and wood stud walls [4].

Table 4. Typical SIP Whole Wall R-values

Thickness	EPS	XPS	PUR
Density in Panel (lb/ft³)	0.90	1.5	2.3-2.5
4-1/2"	13.1	17.7	22.7
6-1/2"	19.9	27.2	35.1
8-1/4"	26.0	35.5	46.0
10-1/4"	32.9	45.0	NA
12-1/4"	39.8	54.6	NA

4.2   Moisture Protection

Since the SIPs' foam core acts as a vapor barrier, the weather barrier must be permeable to allow the SIP's sheathing panels to dry outward. A continuous air space, between the drainage plane and the exterior cladding, and vented openings at the top and bottom of the walls to allow for convective airflow is recommended to ensure adequate drying of the SIPs. This also applies to SIPs used as the roof structure. Air should be able to flow under the roofing material between the eave and the ridge. In addition, all panel joints, openings around windows and doors, and other chases should be properly sealed and/or flashed to prevent moisture infiltration. Particular attention to detail that ensures that interior air infiltration never reaches the outer sheathing layer is imperative. For areas subject to flooding, waterproof sheathing materials such as cementitious skins or thermoplastic skins are an ideal alternative to OSB. However, if SIPs with OSB sheathing come into contact with water, the structural integrity of the panels can be saved if the OSB is quickly exposed to allow drying [5].

4.3   Fire Safety

Since the majority of SIPs construction is for Type V construction where the SIP walls are load-bearing, NFPA 285 compliance does not apply. At this time, there appear to be no NFPA 285 tests that have been performed for SIPs wall construction. Refer to a Building Envelope Consultant if you plan to use SIPs construction where the NFPA 285 test may be required.

4.4   Acoustics

SIPs insulate against high-frequency noise better than low-frequency noise. SIPs are not recommended to use as floors over an open interior space without the application of a sound barrier.

4.5   Material and Finish Durability

The fastener requirements for exterior cladding and interior finishes are specific to the panel manufacturer; consult the manufacturer's specifications for this information. It is recommended that a ventilation space is created with furring strips between the exterior face of the panel and the exterior cladding. This allows for the panels to dry out when water vapor enters the panel.

4.6   Maintainability

The quality of SIPs is set in the manufacturing stage. Proper lamination and smooth surfaces and edges will ensure that the SIPs can endure long-term use as long as the structural skins are properly protected from degradation. It is important to note that if moisture causes deterioration of the skin, then there is a structural issue that must be repaired. Repairs can require the replacement of a much larger area than just the deteriorated portion. Foam insulation is subject to insect and rodent infestation. Insecticides are added to the panels during manufacturing or later on-site.

5.   General Design Details of Structural Insulated Panels

The SIP exterior design building envelope form is only limited to the design imagination (Figure 9). The SIPs can accept any type of properly designed exterior cladding. The SIP panel joints, voids, and penetrations are to be provided with an airtight seal by continuous foam sealant, gaskets, and SIP tapes (Figure 10). Continuity of the inner air seal is imperative to long-term performance. SIP span and structural limit requirements should be verified. SIP nail, screw, and Cams - attachment patterns, fastener types, and spacing requirements should be verified (Figure 11 and Figure 12).

Figure 9. SIPs used for wall and roof panels in a complicated architectural form

Figure 10. SIPs with drywall on a timber frame structure

Figure 11. SIPs for wall panels with drywall and roof panels with tongue and groove

Figure 12. SIPs showing microlam and truss hanger connections, strapping and wiring

6. General Assembly Principles of Structural Insulated Panels

Following are the general assembly principles and installation considerations:

a) Provide exterior roof and wall ventilation/drainage plane systems.

b) Don't provide plumbing in the exterior SIP walls.

c) Coordinate any in SIP electrical.

d) Properly sealed SIPs will provide for the air, vapor, and thermal barriers.

e) Detail for a continuous inner line of redundant air seal at all joints and penetrations using sealant, foams, tapes, and gaskets.

f) Provide for the exterior wall and roof (WRB) water resistance barriers. Note that the WRB should be vapor permeable and must make all joints water and airtight.

g) Provide appropriate flashing systems at all exterior building envelope openings and penetrations.

h) Properly designed HVAC systems are required to address the air tightness and energy efficiency inherent in SIP-designed buildings.

i) Foundation and/or floor deck to be square and level within tight tolerances for effective SIP installation.

j) Detail shop drawings shall be provided by the manufacturer for coordination and to address the General Detail Principles as mentioned above.

k) SIP panel skins need to have solid full bearing support. Review the installation of the SIP bearing plates for this support.

l) The project design team should review any field cutting of SIPs.

m) Foam sealing of SIP panel joints shall be reviewed for continuous full deep sealing. Usually, proper foam sealant installation can be observed by foam seeping at the joints that will need to be cleaned off the panel exterior surface.

n) The inner redundant air seal is commonly accomplished with gaskets placed over the bearing points, spray foam, and with tapes at exposed joints. Carefully select tapes and primers suitable for panel type for long-term adhesion to the panels. Note that OSB is particularly problematic for most common construction tapes.

7. Building Applications of Structural Insulated Panels

a) SIPs are used for commercial buildings that can have higher temperature levels due to the presence of machines and equipment. These panels provide efficient control of indoor temperature during these conditions.

b) SIPs provide the necessary temperature for the storage of medicines and medical items, especially in pharmacies. These panels are used for refrigerated trucks for transporting drugs and medical items that are temperature-sensitive.

c) SIPs are used for large-scale cooling purposes for coating refrigerators and walk-in freezers to maintain the needful temperature. Hospitals, schools, restaurants, camps, etc., use these facilities.

d) SIPs are used in warehouses to maintain interior cold temperatures. It is suitable for warehouses that store food items, electronic products, and other temperature-sensitive items.

Conclusions

The key benefits of building with SIPs are that they provide healthier air quality. SIPs are highly energy-efficient and have high design flexibility. They provide high strength and thermal performance. Sips can be fast fabricated and installed. They use less energy and reduce waste. However, there is some evidence that interior air infiltration through the joints in the SIP roof panels indicated premature deterioration of the top of the OSB skin of the roof panel joints. The moisture damage was due to evidence of the lack of proper joint panel sealing. Although the total time for manufacturing and assembling a SIPs structure is less than that of a framed structure, more time is required in planning. Openings in the panels, non-orthogonal designs, and electrical, and air ventilation coordination must be determined before the manufacturing of the SIPs. The window installation is similar to that of wood frame construction. The manufacturer's specifications should be consulted to insure proper installation. A properly built SIPs structure will be airtight; therefore fresh-air ventilation is required of the mechanical system to prevent interior moisture problems and the build-up of indoor air pollutants. The contractor and the installers should be experienced with SIPs.

References

[1]   Information on https://theconstructor.org/building/building-material/structural-insulated-panels/554863/

[2]   Information on https://www.wbdg.org/resources/structural-insulated-panels-sips

[3]   M. Morley, C. T. Newtown, *Building with Structural Insulated Panels,* The Taunton Press, 2000.

[4]   A. K. Juneau, J. Lstiburek, SIPA Technical Report- Roof Issue, Structural Insulated Panel Association, 2018, USA.

[5]   N. Uddin, A. Vaidya, U. Vaidya, Thermoplastic Composite Structural Insulated Panels (CSIPS) for Disaster Mitigation Construction, American Society of Civil Engineers, Virginia, USA, 2009.

CHAPTER 7

Glass Fibre Reinforced Gypsum (GFRG) Panels

1.   Introduction

Glass Fibre Reinforced Gypsum (GFRG) Panel, commonly known as Rapid Wall, is a building panel product used in large-scale building construction that is formed of calcined gypsum plaster and reinforced with glass fibres. The panel has voids that may be left empty, partially filled, or filled with reinforced concrete in accordance with structural requirements. It was manufactured to a thickness of 124mm under carefully monitored conditions and has a length of 12m and a height of 3m. The strength of GFRG panels filled with reinforced concrete has been proven through practical tests and research in Australia, China, and India. These panels can be used as load-bearing sections as well as shear walls that can endure lateral loads caused by earthquakes and wind. Glass Fibre Reinforced Gypsum (GFRG) panels can also be used profitably as in-fills (non-load bearing) in conjunction with RCC framed columns and beams in traditional framed construction of a multi-story building, and there are no restrictions on the number of floors. As floor or roof slabs, you can utilise RCC screed and micro-beams (which work as T-beams). With panels of 150mm and 200mm thickness, such buildings can be designed to be load-bearing up to 10 stories in low seismic zones [1]. The building industry uses the prefabricated GFRG panel to build living enclosures for residential, commercial, and industrial projects. It is manufactured in factories. The mechanically fabricated 124mm thick hollow-core panels are made of glass-fiber-reinforced gypsum plaster. A typical cross-section of a wall panel is shown in Figure 1. The look of water-resistant GFRG panels is identical to that of regular GFRG panels. However, the water-resistant GFRG's constituents have been altered expressly to offer water resistance whether applied outside or in moist environments like bathrooms or laundry rooms, etc.

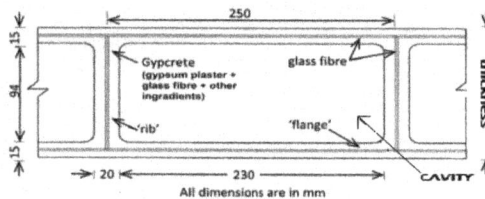

Figure 1. Plan details of a typical cell

2.   Grade and Type of GFRG Panels

The GFRG panels are supplied in the following three grades:

2.1 Class 1: Water Resistant grade of GFRG Panels

The water-resistant grade panels that can be used as floor and wall formwork for concrete infill, external walls, and damp regions.

2.2 Class 2: General grade of GFRG Panels

The common grade panels that can be utilized in dry environments for both structural and non-structural purposes. Typically, these panels are not appropriate for use as wall or floor formwork.

2.3 Class 3: Partition grade of GFRG Panels

Only non-structural internal partition walls in dry locations may be constructed using partition grade panels.

3. GFRG Panels Applications in the Building Construction System

The panel may be used generally in the following ways:

a) As a light, load-bearing walling in one- to two-story buildings. It is possible to use a non-structural core filler material inside the panel without compromising structural integrity, such as insulation, sand, polyurethane, or lightweight concrete..

b) The panel core must be filled with reinforced concrete that is adequately designed to endure the combined impact of lateral and gravity loads since multi-story construction necessitates high capacity vertical and shear load-bearing structural walling.

c) As a partition infill wall in a multi-story frame building, where the panel may also be filled with concrete.

d) As Horizontal floor/roof slabs with screed for T-beam action and reinforced concrete micro-beams.

e) As compound walls and cladding for commercial buildings, as well as pitched (sloped) roofing.

4. Technical Specifications of GFRG Panels

The following raw materials are used in the manufacturing of GFRG Panels:

a) Phosphogypsum: It shall be having purity greater than 90% as $CaSO_4$

b) Glass Roving: The E glass shall be having purity greater than 98%

c) Ammonium Carbonate: It shall be having purity greater than 99.14% as NH_4CO_3

Table 1 shows the mechanical properties of unfilled GFRG panels [2].

Table 1. The mechanical properties of unfilled GFRG panels

Mechanical Properties	Nominal values
Unit weight	0.433 kN/m²
Modulus of elasticity	7500 N/mm²
Uni-axial compressive strength	160 kN/m (4.77 mPa)
Uni-axial tensile strength	34 – 37 kN/m
Ultimate shear strength	21.6 kN/m
Coefficient of thermal expansion	12x10-6mm/mm/°C
Water absorption	1.0% : 1 hr & 3.85% : 24 hrs
Fire resistance: Structural adequacy / insulation	140/140/140 minutes
Thermal conductivity	0.617
U Value	2.85 W/M²K
Sound transmission class (STC) ISO 10140-3:2010	40 dB

5.   Applications of GFRG Panels in Building Construction

The various applications of GFRG panels in Building Construction are discussed as below.

5.1   Rapidwall for Rapid Construction

A trained structural engineer designed the building's construction based on the Design Manual given by the Manufacturer. Using an automated cutting saw, each wall panel is cut at the factory in line with the building plan and design. In accordance with the building drawing, each floor's panels must also have apertures for a door, a window, a ventilator, an air conditioner, etc. For truck delivery to the construction site, panels are vertically placed at the plant into stillages. The stillages are positioned at the construction site adjacent to the foundation for erection using a crane with the required boom length for the construction of low, medium, and high rise buildings. The Reinforced Cement Concrete (RCC) plinth beam is covered with panels, and concrete is poured in from the top. All of the panels are then built in accordance with the building plan after the notation. Each panel is erected level, plumb, and supported by lateral props to keep it there and to keep it securely in place [3]. By cutting open the external flange, embedded RCC lintels are given wherever they are needed. The necessary shuttering and support are given along with reinforcement for lintels and RCC sunshades (Figure 2).

Figure 2. Erection of Wall Panels with Temporary Supports

5.2 Concrete Infill

Using a tiny hose that extends at least 1.5 to 2 metres into the cavities, concrete with 12mm aggregate is immediately pumped into them from a ready-mixed concrete truck. After installing vertical steel reinforcement in accordance with the structural plan and installing clamps at the corners of the walls to maintain the wall panels' precise alignment, this is done. Concrete can be physically poured in modest building construction projects using a funnel. Three layers of concrete, each one metre high, are poured into the panels, with an hour gap between levels. Gravitational pressure causes the concrete inside the watertight compartments to self-compact, eliminating the need for a vibrator.

5.3 RCC Tie Beam All-Around at Each Roof Slab Level

An embedded Reinforced Cement Concrete (RCC) tie beam is specified as a requirement at each floor slab level. When placing horizontal reinforcement using stirrups, the web portion is cut and removed to the required beam depth at the top, and the stirrups are then restored once the concrete has been poured.

5.4 GFRG Panel for Floor/Roof Slab in Combination with RCC

The GFRG panel for the floor/roof slab must be cut to the necessary size and notated. Concrete is poured into the wall joints first, then into additional crevices, cavities, and tie beams made of horizontal Reinforced Cement Concrete (RCC). After a hardwood board between 0.3 and 0.45 metres wide is provided to room span between the walls with support wherever embedded micro beams are present, the roof panels are then raised using a crane. At least a 40mm space separates each roof panel when they are positioned over the wall. This would enable the continuous installation of vertical rods from floor to floor and the construction of a sturdy RCC frames inside the Rapid wall. When embedded micro-beams are present, the top flanges of roof panels are cut to leave at least a 25mm protrusion. Concrete is then poured for the micro-beams and RCC slab once the reinforcement and weld mesh have been installed. The graphical representation of the floor/roof slab system is shown in Figure 3. Figures 4 and 5 show panels with reinforcement being installed in micro-beams and a GFRG floor slab prepared for screed, respectively.

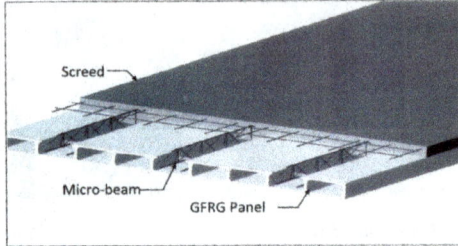

Figure 3. GFRG Roof Panel showing micro-beams, reinforcement and screed

Figure 4. GFRG Roof panels and the reinforcement being laid in micro-beams

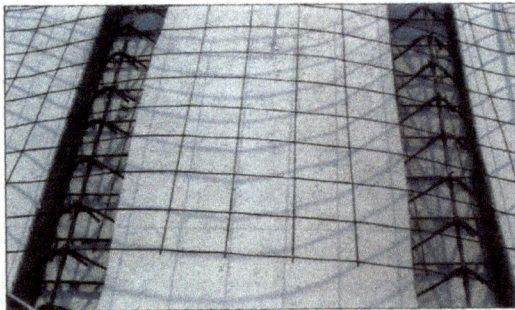

Figure 5. A GFRG floor slab with micro-beam ready for screed

5.5   Erection of Wall Panel and Floor Slab for Upper Floor

The vertical reinforcement of the floor below is given additional length to protrude to 0.45m in order to serve as beginning rods and lap length for the top level. Vertical reinforcing rods, door/window frames, and RCC lintels are cast after the upper level wall panels are constructed. After that, joints are filled and concrete is applied where it is needed. Following that, all of the RCC tie beams are concreted (Figure 6).

Figure 6. Erection of Panels at First Floor

5.6   Finishing Work

The wooden planks and support slabs are removed following the fourth day of concreting the bottom floor roof slab. Experienced Plaster of Paris (POP) plasterers must use wall putty to finish the interior walls and corners of the ceiling. Each higher story also undergoes simultaneous electrical, plumbing, and sanitary work, as well as floor tiling, mosaic or marble work, stair work, etc. Before pouring concrete, conduits can be installed in slabs and walls.

6.   GFRG Panel Selection and Erection

Depending on how the panels are used, different GFRG panel erection techniques must be used. The method of erection of the panel is as follows:

a) Mark the location of the wall's alignment by using line thread.

b) After that, secure the standard door frame by plumbing the wall. For each panel, two holdfasts are needed.

c) Cut the pocket for the electrical points and conduits that will be put within the Rapid wall cavity at the same time.

d) After that, build the panel by using props to support it.

e) Repair switch boxes for electricity.

f) Attach more panels using the same technique as in step one up to the necessary length.

g) Inspect the wall's plumb and line, and then pour concrete into the holdfast gap.

h) Fibre tape must be secured with paint to complete the joints of two panels:

- Create a slot that is 8 mm wide and 2 mm deep at the panel joint.

- Repair the jointing using fibre tape, then paint the surface with stucco.

- The joints between a Rapid wall and an RCC column or beam shall be finished with stucco paint reinforced with reinforcing fibre from recycled cement bags..

- Use stucco paint to fill in the spaces around electrical outlets and between walls and slabs or beams.

7.   Glass Fibre Reinforced Gypsum for Load-Bearing Structural Walling

Reinforced concrete filler or R.C. beams are used to connect cross walls, the floor, the roof, and the multi-story construction's conventional use of Glass Fibre Reinforced Gypsum (GFRG) as load-bearing structural walling. All GFRG wall panels must be installed on a system of RC plinth beams that are properly supported by the ground (Figure 7). "Starter bars" must be embedded in the RC plinth beams at the critical locations where the cavities are to be filled with reinforced concrete, with the appropriate lap length. This ensures that the superstructure and foundation are connected at the ground level over the length of the wall and across the network of RC plinth beams [4].

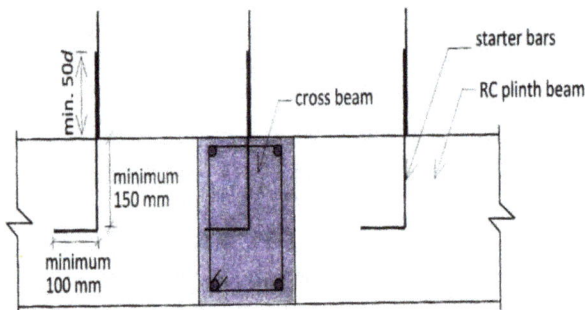

Figure 7. The starter bars in RCC plinth beam for erection of GFRG panels

By employing the proper detailing (insertion of starting bars with anchorage), it is possible to link the GFRG wall and the existing floor when adding a new GFRG floor on top of an existing RC structure, as shown in Figure 8.

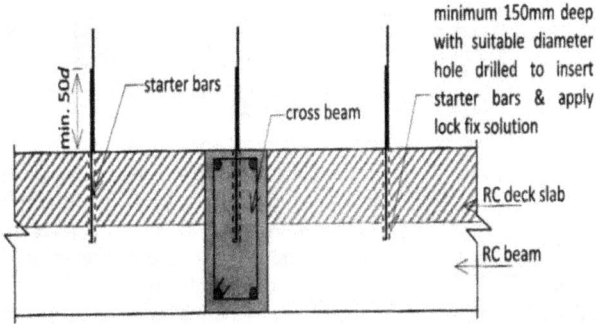

Figure 8. Details of connection between existing floor and GFRG walls

A horizontal RC tie beam that is embedded and placed on top of each wall must be provided when GFRG panels are utilized as structural walling. Cutting and removing the top portion of the GFRG panel web as depicted in Figure 9 leads to the suggestion that the tie beam should be 200mm deep and 94mm wide.

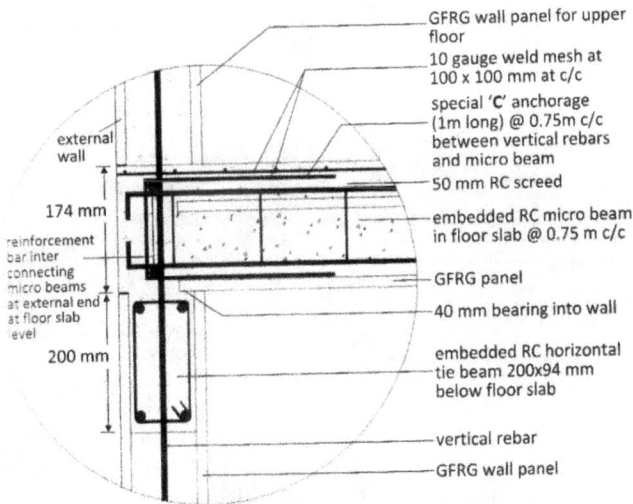

Figure 9. Details of R.C. tile beam and wall junction

8. Glass Fibre Reinforced Gypsum as Floor / Roof Slab

In addition to RCC, GFRG panels can be used for intermediate floor slabs and roof slabs. The strength of GFRG slabs can be significantly increased by using reinforced concrete micro-beams. To supply embedded micro-beams, the top flange of the relevant cavity is cut and removed, leaving a minimum 25mm flange protruding from both ends. An RC concrete screed with a minimum thickness of 50mm is placed over the GFRG floor panel, which is reinforced with weld mesh at a minimum size of 10 gauges, 100 mm x 100 mm. The RC screed and micro beam work as a set of embedded T-beams together. The thickness of the RC screed, the reinforcement, and the distance between the RC micro beams are all determined by the span and magnitude of the load. The horizontal tie beam, integrated RC micro-beams, concrete screed, and vertical rods in the GFRG wall are all coupled in a way that perfectly secures the floor/roof slab and walling system. Gypsum wall panels with glass fibre reinforcement as well as slabs in columns and beams. GFRG panels can also be used in high-rise structures. An example of high-rise building design combining RC columns and beams and GFRG composite construction is shown in Figure 10. The vertical wall panel above beam must have a clearance margin from the edge of the RC beams, as shown by the dotted lines on the beams.

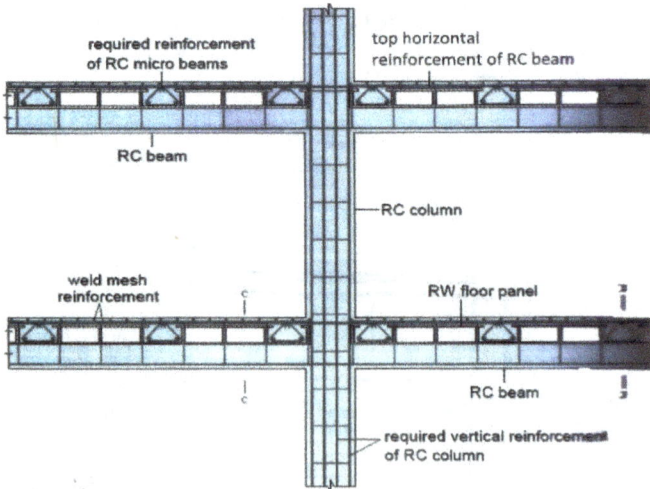

Figure 10. Columns and beams with GFRG composite construction

Figure 11 to Figure 15 shows the joinery details of various structural components with GFRG composite construction methods for tall buildings.

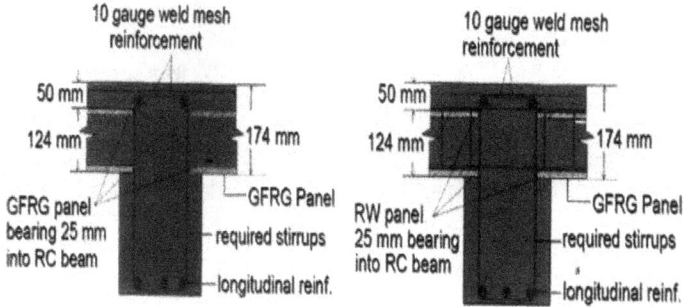

Figure 11. RC Columns and T beams with GFRG composite construction details

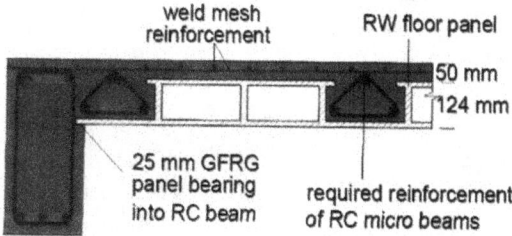

Figure 12. Sectional details of external beams with floor slab across RC micro-beams

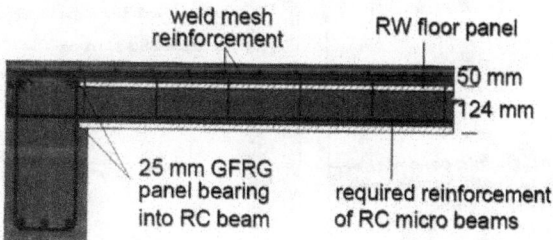

Figure 13. Section details of external beams with slab and longitudinal of micro-beams

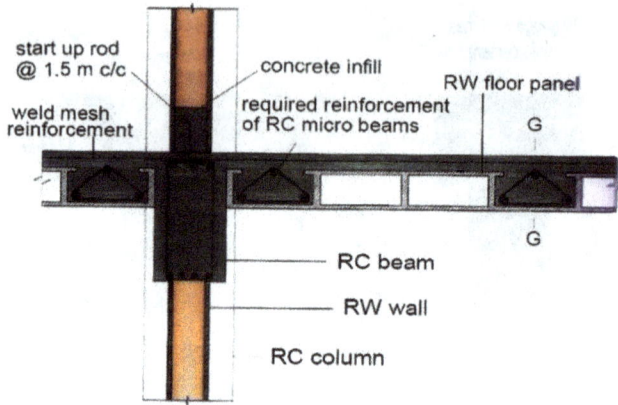

Figure 14. Sectional details of embedded micro-beams

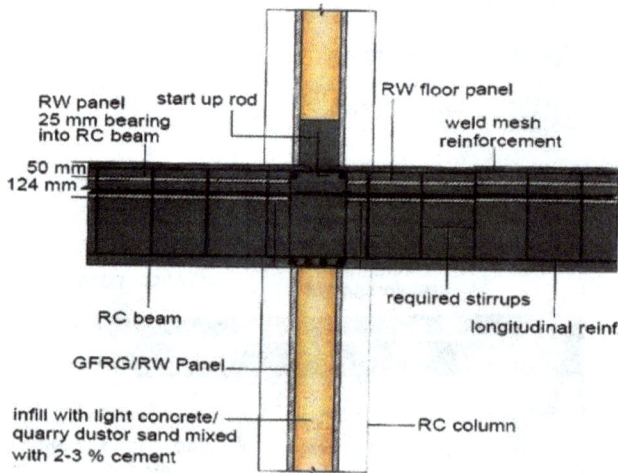

Figure 15. Section through micro-beams and slab beam junction

9. Advantages of Glass Fibre Reinforced Gypsum (GFRG) Panel System

The building's structural weight is significantly reduced as a result of its modest weight. These panels weigh only 43 kg/m², which is quite little. Even after part of the voids have been filled with concrete, the overall weight of the building is still significantly lower, which helps to significantly reduce the seismic stresses in the design and reduce foundation and overall building costs, especially for multi-story buildings. These panels can be used for floors, roofs, staircases, and walls in addition to walls. Building construction proceeds more quickly as a result. In comparison to conventional building, the construction of a building can be completed quite quickly by using these GFRG panel systems. It needs less labour and workmanship. Because the panels are just 124mm thick, they have a substantially smaller footprint than conventional buildings for the same carpet area [2]. This is especially beneficial for multi-story mass housing. Given that it is produced using leftover gypsum, it is a green building material. Gypsum from industrial waste is used in it. For walls and ceilings, there is no plastering necessary. uses a lot less water, steel, cement, and sand than traditional construction. It has a lower carbon footprint and uses a lot less embodied energy. By utilizing less scarce natural resources, such as cement, steel, sand, burned clay bricks, and concrete blocks, these glass fibre reinforced gypsum panels preserve energy. Figure 16 shows the staff housing quarters constructed with the help of GFRG panels at IIT Madras, India

Figure 16. Staff housing quarters constructed by GFRG panels at IIT Madras, India

10. Limitation and Disadvantages of GFRG Panels

For walls with circular or greater curvature, the GFRG Panel System cannot be used. For residential buildings, the clear span must not exceed 5 metres; for non-residential buildings, it must not exceed the limits outlined in the Design Manual. On-site cutting of the GFRG Panel requires specialized equipment. A crane needs extra room to manoeuvre during the construction phase. Due to the sophisticated GFRG Panel design, the building approach is less cost-effective. Labourers with a high level of experience and competence are required to install the GFRG panels. The GFRG Panel should be handled carefully when being transported and erected.

Conclusions

As long as the construction detailing necessary to deal with shrinkage or expansion is known, the Glass Fibre Reinforced Gypsum Panels are a great building material. A more contemporary approach to building a walling system than the more conventional brick masonry techniques is the use of GFRG panels. In particular, the engineering viability and characteristics of GFRG panels as a walling material in building construction are covered in this study. In many ways, these panels function better than burned clay brick masonry. The production of GFRG panels from the raw material of gypsum, specifically natural gypsum, mineral gypsum, phosphor gypsum or chemical gypsum, with a purity of more than 90%, suggests less energy use when compared to traditional building materials, such as bricks, concrete, etc. Due to the better efficiency of the carpet area and shorter constructing time, the GFRG system has been demonstrated to be highly advantageous in all elements of construction, particularly in time and money. If future study looks at the risk analysis for using the system in repetitive projects, the owner and designer will be able to estimate the cost for the GFRG system during the design process. It is necessary to investigate novel methods for GFRG system cost reduction while maintaining system performance. Future studies on the environmental performance of places utilizing the GFRG panel technology can take the environmental issues into consideration. Additional research might focus on the quality controls for on-site GFRG system installation inspection.

References

[1] P. K. Adlakha, New Building Materials, and Technologies, Vol. IV, Indian Building Congress, New Delhi, India, 2019.

[2] BMTPC, GFRG / Rapidwall Building Structural Design Manual, IIT Madras and Building Materials and Technology Promotion Council, New Delhi, India, 2014.

[3] M. Janardhana, Cyclic behaviour of glass fibre reinforced gypsum wall panels, Unpublished Ph. D. Thesis, Indian Institute of Technology Madras, Chennai, India, 2013.

[4] S, Geethu, R. Renjith, Formulation of an Alternate Light Weight Concrete Mix for Concrete Filled Glass Fibre Reinforced Gypsum (GFRG) Panels, International Journal of Science and Research, 4(7), 2015.

Novel Approaches in Materials and Technologies for Building Construction   Materials Research Forum LLC
Materials Research Foundations **158** (2024)   https://doi.org/10.21741/9781644902936

CHAPTER 8

Monolithic Concrete Construction Technology

1. Introduction

The Monolithic Concrete Technology is a rapid and disaster resistant construction with aluminum formwork is an upcoming technology which is enabling the cost-effective and rapid mass housing projects. The monolithic concrete construction technology is also sometimes named shear wall technology. In this system, instead of conventional R.C.C. framed constructions of column and beam, all walls, floors, slabs, columns, beams, stairs, together with door and window openings are cast in place monolithic using an appropriate grade of concrete in one operation at the site by the use of specially designed, easy to handle (with minimum labor and without the use of any elaborate equipment) modular formwork made of Aluminium/aluminum-plastic composite. Using the special formwork system, rapid construction of multiple units of repetitive type can be achieved. The entire operation essentially comprises fitting and erecting the portion of shuttering as already determined, (the optimization in use as determined by appropriate planning) placement of reinforcement, and then carrying out concreting of the walls and slab simultaneously. Props are so designed that they stay in position while de-shuttering of slabs. The cycle time of construction is 3-5 days per floor [1].

Monolithic construction is a method by which walls and slabs are constructed together. In this method, fresh cement concrete is poured in lightweight aluminum formwork system having required reinforcement bars for needed strength. As the walls and slabs are cast in one go, the operation is very fast. This is ideal for multi storied construction, allowing speedy construction on mass scale. This technology offers speedier solutions to rapidly increasing housing shortage in urban areas by optimal use of time, money and building materials like steel and cement. It promises accelerated construction at optimized cost and time when we go for mass housing especially for economically weaker sections and low income groups who are large in number without houses. It is a highly efficient technology which facilitates concreting of all the components like walls, roof etc. simultaneously, resulting in a structurally very sound monolithic construction.

Monolithic Reinforced Concrete Construction System uses a formwork system that allows casting walls and slabs according to a pre-defined cycle. It combines the speed; quality and accuracy in production with the flexibility and economy of in-situ construction. The result is a reinforced concrete structure, the surfaces of which are of sufficiently high quality to require only minimal finishing while the end walls and facades are easily completed. Since footing, wall and slab are designed as an integrated unit and Reinforcement is continuous and interlocks with all the structural components of the building viz., foundation, walls and slabs. Hence Columns and beams are not required which leads to Slender sections thereby providing high resistance to earth quake, cyclone, wind and flood [2].

The salient features of this technology have been summarized as follows:

a) The technology requires unskilled and semi-skilled labors (hand-held) and does not require the use of expensive construction equipment. Hence cost effective.

b) The modular design of mass housing schemes provides excellent opportunity for reuse of form work which makes the technology very cost effective.

c) The material of formwork (either aluminum or HDPE) is recyclable and completely eliminates the use of precious natural resource wood. Hence the technology is environmental friendly.

d) This technology helps excellent quality control of the entire structure as per BIS and all the international standards.

e) No need of bricks, blocks and plastering.

f) Because of reduced dead load of about 50%, superstructure and foundation cost can be minimized without compromising on strength.

g) Its box-like behavior gives the building very high structural strength, against vertical and horizontal forces thereby making it earth-quake and wind/cyclone resistant.

h) Excellent finished surface avoids expensive plastering and enhances a relatively water resistant surface.

i) Due to reduced wall thickness for a given Plinth area more Carpet area is available.

j) Monolithic concrete construction helps in precise scheduling and assured quality control.

2.   Basic Material Requirement

a) The formwork system is a propriety system and designed as per the load requirements of the structure. It should have adequate stiffness to weight ratio, yielding minimum deflection under concrete loading. The panel should fix precisely, securely, and require no bracing. IS 14687:1999 Guidelines for false work for concrete do not cover requirements of this special type of formwork system.

b) Concrete shall be of appropriate grade based on the environmental condition as per IS 456:2000

c) Reinforcement shall conform to IS 1786:2008

3.   Details of Formwork

The formwork systems used are made of lightweight Aluminium. The recommended concrete forms generally use a robotics welding system for manufacturing. A soft alloy weld wire is utilized in the concrete form weld process. Fixing of the formwork is done using a tie, pin and wedges system, which does not require very skilled labor to do the job. The hardware consists of hundreds of standard pieces of formwork equipment, manufactured to a fine tolerance. The majority of the equipment comprises panel sections while the rest includes vertical and horizontal corner sections, bulkheads, and special floor slab beams that can be dismantled without disturbing the props supporting the floor slab concrete. Even the larger components are light enough to be handled by a single worker. Consequently, the need for cranes or other heavy handling equipment is eliminated. Steel pins and wedges join all individual pieces of formwork composite. The only tool required in assembly is the regular hammer. It certainly eliminates the need for skilled workers.

The technology is extremely adaptable to any building design. Based on the architectural and structural drawings of the building, a process of computer modulation is carried out. It involves iteration and optimization techniques, which select the most economical and practical fittings of the standard formwork components. Formwork layout and shop drawings are prepared for each project. This is the most important step in the pre-construction process and herein lies the uniqueness of the technology. Due to the fine tolerance achieved in the formwork components, consistent concrete shapes and finishes are obtained floor after floor, building after building, conforming to the most exacting standards of quality and accuracy. This allows plumbing and electrical fittings also to be cast with the certain knowledge that there will be an exact fit when assembled. The dimensional accuracy of the concreted works also results in the consistent fitting of doors & windows. Since the formwork for all components of the structure is erected at one time, placing concrete with suitable slumps is produced by adding superplasticiser and vibrated by high-frequency pokers. Self-compacting concrete can also be used, which eliminates the requirement of vibration. Inserts are provided in the forms at a predetermined location for drainage, water pipelines and electrical cables, which will pass through walls. This eliminates any breaking of the concrete after it has been cast. The formwork can be designed based on the requirements of the project. A repetition of about 500 cycles can be done.

4.   Construction Process

The process of construction generally encompasses:

a) Construction up to plinth level (min. 150mm thick) cast monolithically using specially designed formwork.

b) Tying of vertical and horizontal reinforcement of the walls, manually fixing of cover blocks to maintain the cover and wall thickness.

c) Fixing of M.S. flats with pin and wedge connection to both the wall panels.

d) Erection of channel section for slab above the wall panels.

e) Fixing of joints and lintel members to the wall panels.

f) Fixing of internal conduit wiring etc.

g) Concreting of walls and roofs.

h) De-shuttering and curing.

i) Repetition of the process for the next storey.

5. Durability

As construction is done using concrete, the durability of the structure can be achieved by using proper ingredients, grade of concrete & mix design as per IS 456:2000. The thickness of the wall is a minimum of 100 mm with the reinforcement placed in the middle. Therefore, the adequate cover is likely to be maintained. For walls, more than 125 mm thickness reinforcement is provided on both the faces with cover required as per IS:456 as per climatic conditions and fire resistance.

6. Thermal Behaviour of Structure

A field study of structures with similar orientation constructed with conventional brick masonry and 100 mm thick monolithic concrete revealed that "For 115 mm thick Filler Brick wall if the external temperature is 40°C, the internal temperature at the end of 4 hours will be around 37.8°C. For 100 mm. Monolithic R.C.C. wall, the internal temperature will be around 38.7°C at the end of 4 hours for an external temperature of 40 °C.A 100 mm thick RCC Walls and Roof has a thermal transmittance value of 3.59 W/m^2 K (IS 3792:1978). Since it is more than the normal plastered brick wall (thermal transmittance 2.13 W/m^2 K), it is recommended to ensure proper planning for air ventilation provisions in housing units (IS 3792:1978). Nowadays, special external paints with admixtures are available which are heat resistant and can be used for application on external surfaces to achieve a reduction in inside temperatures. Moreover, other measures such as providing shadows, cladding of the walls facing South-West can also be adopted [3].

7. Acoustical Properties of Monolithic Concrete Construction

The average sound reduction for 100 mm. concrete is $\geq$ 45 db (IS: 1950-1962).

8. Ease of Fixing Services

All electric and plumbing fixtures, lines have to be pre-planned and placed before concreting is done. Post-construction alteration is not desirable. In the Indian conditions, predetermined, grooves in running length during concreting have been tried successfully, wherein the conduits and flexible water pipes can be laid later. The grooves are then filled with mortar with chicken mesh wrapped over the pipes to take care of temperature and shrinkage stresses.

9.   Speed of Construction

The speed of construction mainly depends upon the economy required and the construction period of the project. Various work cycles can be achieved such as 4, 6, 8, or 10-day slab cycles, depending upon the number of sets of formwork.

10.   Finishing

Although the high tolerance finish means that no further plastering is required, yet a coat of 3 to 5 mm thickness as a skin coat of POP / Putty needs to be applied internally to walls before application of paint. For exteriors either cement-based waterproof paint or textured paint can be applied. As the External surface is smooth, a coat of primer/bonding agent enhances the bond of paint with the bare RCC wall.

11.   Scale of Economy

The scale of the economy depends upon the volume of work and the maximum number of repetitions of the formwork achievable for the estimated period of construction. Minimum 100 repetitions of the formwork are desirable. For a small project of less than 500 similar units, this system may not prove to be economical.

12.   Other Features

Pre-designed formwork acts as assembly line production and enables rapid construction of multiple units of repetitive type. With proper planning, a slab cycle of 4 days can be achieved, which reduces the construction time considerably. It is flexible in design and can form any architectural or structural configuration, such as stairs, windows, etc. The formwork is manually handled. There is no need for heavy equipment and cranes etc. The finish is such that it requires no separate plaster.

13.   Comparison of Monolithic construction with Conventional construction

The comparative analysis of Monolithic construction with Conventional construction is shown in the Table 1.

Table 1.Comparison of Monolithic construction with Conventional construction

Conventional	Monolithic
Less Structural Stability against non-gravity loads.	High Structural Stability against non-gravity loads.
Requires more construction activities, time and labor	It is speedy and requires fewer work forces.
Higher construction cost.	Cheaper by 10% than conventional methods. Lower construction cost.
More dead weight. No savings in foundation cost.	Lesser dead weight. Savings in foundation cost.
More number of skilled masons is required.	This technology requires only trained and unskilled workers
No consistency in shuttering. Irregular shape may spoil the final finish.	The Monolithic technology has simple modular light weight formwork system which gives a defined shape.
Wall surfaces are too rough. Inner and outer surfaces require plastering	The form finish concrete surface is smooth. Needs no plastering
Periodic and regular maintenance required.	The cement concrete monolithic structure once built will always be Maintenance free

14. Merits of Monolithic Concrete Technology

Providing housing for the poor, and slum dwellers is a gigantic and a complex problem. The Government needs to adopt the technology that provides solution to meet housing needs of millions of people. To do this, faster construction by launching Mass housing projects with monolithic technique is one of the solutions to the overgrowing problem. The speed & quality construction drives this technology.

a) Cost effective - average 15% cost saving for load bearing wall over conventional timber/plywood shuttering

b) Increased Carpet area/Usable floor space of (5 - 8%) over Conventional Design

c) Speedy construction - 1/5th - 1/6th of time required to complete construction against the conventional

d) Easy to use, since it's simplified design

e) Monolithic construction-box type strong structures with minimum/ no joints 14

f) Structures are better earthquake and wind resistant

g) Highly durable structures - required least maintenance

h) Smooth finish of wall/roof/floor- ready to take the paint

i) Block/Brick masonry eliminated thus Internal & External Plaster eliminated

j) Environment-friendly - no wood/timber used

k) Formwork-Lightweight section are used-hence easy to lift and used

l) Aluminium shuttering material has higher scrap value compared to plywood/steel

15. Limitations and Demerits of Monolithic Concrete Technology

As observed from the field, following drawbacks are noticed:

a) Generally, used only in mass housing project with same plan and the multi-storey structures having same plan area at all floors.

b) The formwork requires careful and gentle handling both during erection as well as during opening, removal, and transportation.

c) Huge initial investment for formwork procurement. The initial investment for the formwork system is high compared to other forms and thus it requires the construction of a minimum of 500 similar houses for the economy.

d) There is no saving in construction cost in a single-storey structure.

e) About 70-80% of formwork elements shall be useful for new project (new plan) after first project, rest shall be suitably designed and procured for next projects

f) A lead time of about 3 months is required for the initiation of work, as the formwork is custom designed and manufactured as per the requirement of the structure.

g) Initial setting of formwork at site as per drawing dimension, may take more time

h) The training of labor to handle designed formworks at site.

i) All the service lines are to be pre-planned.

j) All the utilities have to be laid and embedded in concrete walls/slabs, hence relocation, repair of these will be very difficult. Therefore, the post-construction alterations are difficult.

k) Since, structure is monolithic and wall thickness is less, these houses are little warmer than the conventional during summer. However, this problem can be overcome by using suitable thermal insulated paint for external wall or by adopting simple thermal insulation techniques or growing creepers and plants around the external walls.

16. Visuals of Construction Methods on Site

Figure 1 to Figure 6 shows the various stages of construction and the few buildings constructed with monolithic technology.

Figure 1. Laying RCC strip foundation with dowel bars and wall reinforcement

Figure 2. Formwork for walls below plinth

Figure 3. External wall corner formwork

Figure 4. Formwork and reinforcement for slab ready for casting along with the walls

Figure 5. Casting of staircase monolithically

Figure 6. External wall formwork for first floor structure

Conclusions

The Monolithic Concrete Technology adopted in many projects in India. The slum Housing Project at Sathagally and Rajivnagarin Mysore city has achieved a cost reduction in the construction to the extent of 15-20%. There was no need for plastering and for providing lintels. This technology can be conveniently used for all massive housing programs for poor, the low income groups and luxury housing. The defects that are commonly committed in traditional and conventional construction methods can be avoided in monolithic concrete construction. These structures are efficient against earthquake due to its single rigid block nature. Unlike load bearing construction using brick and stones which are generally weak against disasters, the monolithic technology provides resistance and safety against earthquake and cyclones. The cost of repairs and maintenance would also be reduced as compared to other constructions. However, the thermal discomfort in summer can be prevented by proper thermal insulation through the method of creepers around the walls and roofs, insulation painting etc.

The Conventional and Monolithic Construction technologies are compared in which later is 10 to 15% more economical and 25% faster. Monolithic Construction technology is very suitable for mass housing and rehabilitation housing projects. Most important observation is that all those living in the houses felt that house is comfortable and they were satisfied with the quality. Most of the key components like walls, columns, beams, floor slabs, staircases, balconies, openings, hoods, storage lofts, etc., are monolithically concreted in-situ. No need of bricks, blocks and plastering. Because of reduced dead load, superstructure and foundation cost are minimized without compromising on strength. This type of construction is a fast-track construction [2]. All services like electrical, plumbing and sanitary are embedded before concreting which avoids breaking and making of structures. Whenever there is a challenge of construction in least possible time with best quality.

References

[1]   BMPTC, 'Monolithic construction for Mass housing', BMTPC publication, New Delhi, 2002.

[2]   Information on https://atimysore.gov.in/wp-content/uploads/action-reserach-on-monolithic-concrete-technology.pdf

[3]   P. K. Adlakha, New Building Materials and Technologies, Vol. IV, Compendium of New Building Technologies, Indian Building Congress, New Delhi, India, 2019.

Novel Approaches in Materials and Technologies for Building Construction   Materials Research Forum LLC
Materials Research Foundations **158** (2024)   https://doi.org/10.21741/9781644902936

CHAPTER 9

Formwork System for Building Construction

1. Introduction

With the advancement of science and technology, the construction of high rise buildings became easy by inventing new machinery and new building techniques. One such area related to high-rise construction is the type of formwork used in the construction. Formwork is a mould or die used to shape and support the concrete until it attains sufficient strength to carry its own weight. Formwork which holds and supports wet concrete till such time it cures, is a very vital element in concrete construction. It has become foremost to have speedy construction and timely completion of projects. The formwork is a kind of 'baking tin', which holds the concrete until it hardens to attain the required shape and size. Formwork constitutes approximately 30% of the cost and 60% of the time in concrete construction. The quality of concrete finish and soundness of concrete depends very much on the formwork system [1].

With the advancement of science, man used plywood instead of timber planks and pipe supports with various kinds of jacks instead of timber supports. Traditional formwork for concrete construction consists of bespoke solutions requiring skilled craftsmen. This type of formwork often had poor safety features and gave slow rates of construction on-site and huge levels of waste [2]. But nowadays, modern formwork systems for superstructure construction are commonly adopted. Formwork should be properly designed, fabricated, and erected to receive fresh concrete. If formwork is not done properly the desired shape of concrete is not possible. When concrete is compacted, it exerts pressure and the formwork must be strong and stable to take this pressure. The form should be leak proof to retain the concrete and slurry.

Formwork is a structure, usually temporary, used to contain poured concrete and to mold it to the required dimensions and support until it can support itself. It consists primarily of the face contact material and the bearers that directly support the face contact material. Formwork systems used for concrete frame construction have continued to develop significantly since the early 1990s. The major innovations have focused on on-site efficiency of production, health and safety, and environmental issues, driving the concrete construction industry towards ever-increasing efficiency. Different formwork systems provide a wide range of concrete construction solutions that can be chosen to suit the needs of a particular development. Modern formwork systems are designed for speedy and efficient construction.

Formwork system can be generally classified as Vertical Systems (wall and column) and Horizontal Systems (slab and beam). The material serving as the contact face of forms is known as sheathing and it is used in both the vertical and horizontal systems. This chapter describes generic types of modern formwork system that are widely available globally and considers their applications, advantages and main features related to HSE (Health, Safety and Environment) parameters and sustainability performance. These systems are pre-engineered to provide increased accuracy and minimize waste in construction and most have health and safety features built-in [3].Figure 1 shows a typical formwork for casting a reinforced cement concrete column.

Figure 1. A typical formwork for a reinforced cement concrete column

2.   Requirements of a Good Formwork System

Following are the salient requirements for a good formwork system:

a) The formwork is to be strong enough to withstand the dead and live loads, forces caused by ramming and vibration of concrete and other incidental loads imposed upon it during and after casting of concrete.

b) Suitable arrangements should be there to avoid any settlement in the form work either before or during the placing of concrete.

c) The formwork should be of sufficient stiffness to avoid excessive deflection and joints should be tightly butted to avoid leakage of cement slurry.

d) The formwork must be accurately set out so that the resulting concrete product is in a right place and is of correct shape and dimensions.

e) The formwork surface is to be coated with suitable mould oil so that good concrete quality and surface finish can be achieved.

f) The form panels and units should be so designed that their maximum size does not exceed and can be easily handled by hand or mechanical means. The formwork should be designed and constructed to facilitate for adjustments, leveling, easing and striking without damage to the formwork or concrete.

g) It should be ensured that the optimum stock of formwork for the size of work force, the specified time schedule and flow of materials.

3.   Economy in Formwork

The following points are to be kept in view to effect economy in the cost of formwork:

a) The plan of the building should imply a minimum number of variations in the size of rooms, floor area, etc. to permit reuse of the formwork repeatedly.

b) Design should be perfect to use slender sections only in the most economical way.

c) Minimum sawing and cutting of wooden pieces should be made to enable reuse of the material many times. The quantity of surface finish depends on the quality of the formwork.

4.   Components of Formwork

There are many different components of Formwork system. Figure 2 shows the different components of a typical slab formwork system. They are summarized as below:

4.1   Propping and Centering

The props used for centering may be of steel, aluminum, timber post.

4.2   Shuttering

The shuttering can be made up of timber planks or it may be in the form of panel unit made either by fixing ply wood to timber frames or by welding steel plates to angle framing.

4.3   Provision of Camber

Certain amount of deflection in structure is unavoidable. It is therefore desirable to give an upward camber in the horizontal member of concrete structure to counteract the effect of deflection.

4.4   Surface Treatment

The formwork should be cleaned of all rubbish particularly the dust & chippings etc. Before laying concrete the face of formwork in contact with concrete shall be cleaned and treated with release agent like raw linseed oil or soft soap solution as to prevent the concrete getting struck to the formwork.

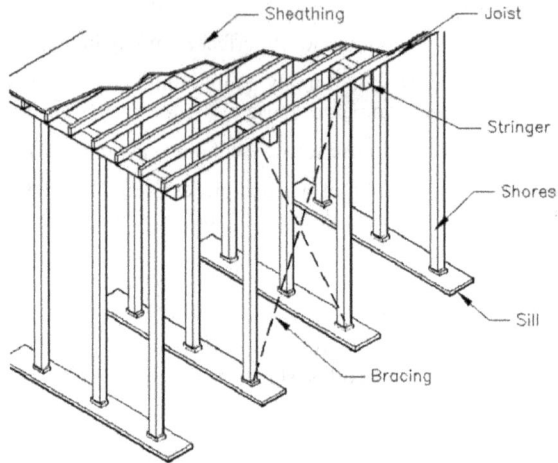

Figure 2. The different components of a typical slab formwork system

5. Design and Layout of Formwork

The design and construction of formwork shall take account of safety and of the surface finish required. The formwork shall be sufficiently rigid and tight to prevent loss of cement slurry from the fresh concrete or the formation of fins and honey combing on surface. For further details regarding design, detailing, etc. reference may be made to IS: 14687. The formwork and its supports maintain their correct shapes and profile so that the final concrete structure is within the limits of the specified dimensional tolerances. They shall be designed to withstand the combination of self-weight, reinforcement weight, wet weight of concrete, concrete pressure, construction and weather loads, together with all incidental dynamic effects caused by placing, vibrating and compacting the concrete. The formwork shall never be tied to or supported by the reinforcement steel.

All horizontal and vertical formwork joints on exposed surface shall be so arranged that joint lines are staggered. Where the contractor proposes to make up the formwork from standard sized manufactured formwork panels, the size of such panels shall be approved by the Engineer in charge before they are used in the construction of the works. The finished appearance of the entire elevation of the structure and adjoining structures shall be considered when planning the pattern of joint lines caused by formwork and by construction joints to ensure continuity of horizontal and vertical lines [3].

6. Need for Modern Formwork System

The earliest formwork systems made use of wooden scantlings and timber runners as it enabled easy forming and making at site. But these wooden scantlings and timber runners tend to lose their structural and dimensional properties over a period of time and after repeated usage thus posing safety problems. Many of the accidents take place in Reinforced Cement Concrete (RCC) construction because of inferior formwork and scaffolding.

The modern formwork systems are designed for speedy and efficient construction. They are designed to provide increased accuracy and minimize waste in construction and most have enhanced health and safety features built-in.

7. Different Types of Formwork systems

The main types of formwork systems in use now are described as follows:

7.1 Traditional Timber Formwork

The traditional slab formwork technique consists of supports out of lumber or young tree trunks that support rows of stringers assembled roughly 3 to 6 feet or 1 to 2 metres apart, depending on thickness of slab. Between these stringers, joists are positioned roughly 12 inches; 30 centimeters apart upon which boards or plywood are placed. The stringers and joists are usually 4 by 4 inch or 4 by 6 inch lumber. The most common imperial plywood thickness is ¾ inch and the most common metric thickness is 18 mm. The formwork is built on site out of timber and plywood or moisture-resistant particleboard. It is easy to produce but time-consuming for larger structures, and the plywood facing has a relatively short lifespan. It is still used extensively where the labor costs are lower than the costs for procuring reusable formwork (Figure 3). It is also the most flexible type of formwork, even where other systems are in use, complicated sections may use it. Figure 4 shows the timber Formwork for casting of a concrete wall.

Figure 3. Timber Formwork for concrete column

Figure 4. Timber Formwork for casting of a concrete wall

7.2   Plywood Formwork

Plywood formwork is a lightweight, durable and eco-friendly system that provides high quality concrete cast surface. It is prepared by combining the specially prepared three layers on top of each other. Plywood formwork is applicable wherein a large surface needs be coated with a light and durable material. It is especially preferred in molds prepared for pouring concrete in constructions (Figure 5). Usually, plywood boards are made of birch, poplar, spruce, or pine but there are many types of materials used by plywood manufacturers in different parts of the world. While the type of plywood used is important, the real opportunity for making a difference for the final concrete result is the type of coating: the film that is applied on the plywood surface [4].

Resin-bonded plywood sheets are attached to timber frames to make up panels of the required sizes. The cost of plywood formwork compares favorably with that of timber shuttering, and it may even prove cheaper in some instances given the following considerations:

a) It is possible to have a smooth finish in which case on cost in surface finishing is there.

b) By the use of large-size panels, it is possible to affect saving in the labor cost of fixing and dismantling.

c) The number of reuses is more as compared with timber shuttering. For estimation purposes, the number of reuses can be taken as 20 to 25.

Figure 5. Plywood boards used as formwork for roof slab construction

Advantages of plywood formwork

Plywood has the natural property to absorb moisture, resulting in a sponge effect. This causes movement of the panels, affects the weight and the mechanical properties of the material, ultimately influencing its life cycle[4]. The advantages of plywood formwork are as follows:

    a) Provides high-quality concrete cast surface

    b) Favorable for weight to strength ratio

    c) Minimal deflection due to inherent stiffness

    d) High durability

    e) Can be reused

    f) The panel shear of plywood is nearly double that of solid timber due to its cross laminated structure.

    g) Easy to machine and fix using standard machining and fastening methods

    h) Easy to handle on-site

    i) Good chemical resistance

    j) Wide variety of different overlays and sizes available

    k) Plywood has ecological importance.

7.3   Steel Formwork

This consists of panels fabricated out of thin steel plates stiffened along the edges by small steel angles. The panel units can be held together through the use of suitable clamps or bolts and nuts.The panels can be fabricated in large numbers in any desired modular shape or size. Steel forms are largely used in large projects or in a situation where large number reuses of the shuttering is possible. This type of shutter is considered most suitable for circular or curved structures [5]. Figure 6 shows the steel as formwork for casting of column and wall.

Comparison between Steel and Timber Formwork

a) Steel forms are stronger, durable, and have a longer life than timber formwork and their reuses are more in number.

b) Steel forms can be installed and dismantled with greater ease and speed.

c) The quality of exposed concrete surface by using steel forms is good and such surfaces need no further treatment.

d) Steel formwork does not absorb moisture from concrete.

e) Steel formwork does not shrink or warp.

Figure 6. Steel sheets used as formwork for casting of column and wall

7.4   Aluminum formwork

Aluminum formwork system is almost similar to those made of steel. Aluminum forms are lighter than steel forms due to low density and this is their primary advantage when compared to steel. The shuttering is economical if large numbers of repeating usage are made in construction. The disadvantage is that no alteration is possible once the formwork is constructed [6]. Aluminum Formwork System is a comparatively a new technology, saves cost, time and improves the quality of construction. For repetition of building layouts and for above-the-plinth work, Aluminum Formwork system is very cost effective. Aluminum Formwork panels can be designed for any condition/component of building such as bay windows, stairs, balconies and special architectural features (Figure 7).

Figure 7. Aluminium formwork system for columns beams and slab

7.5   Engineered Formwork System

This formwork is built out of prefabricated modules with a metal frame (usually steel or aluminum) and covered on the application (concrete) side with material having the wanted surface structure steel, aluminum, timber, etc.) as shown in Figure 8. The two major advantages of formwork systems, compared to traditional timber formwork, are speed of construction (modular systems pin, clip, or screw together quickly) and lower life-cycle costs (barring major force, the frame is almost indestructible, while the covering is made of wood; may have to be replaced after a few - or a few dozen - uses, but if the covering is made with steel or Aluminium the form can achieve up to two thousand uses depending on care and the applications).

Figure 8. An engineered formwork system

7.6   Re-usable plastic formwork

The plastic formwork is manufactured from specific grade plastic that neither placed materials adhered to it nor chemical reactions occur between poured materials and the plastic form and consequently there will be no any patch on the surface of reinforced concrete element. Water or freshly placed concrete leakages from different parts of the form is avoided because of the perfect fitness of various parts of the system. Added to that, it is most labor friendly system because not only does it fit and plugged easily but also it is considerably light weight compare with other types of formwork systems. Figure 9 shows the Plastic formwork system for wall construction

Figure 9. Plastic formwork system for wall construction

The nailing and oiling plates are not required prior to concreting because of auto leveling of plugging system that make plates level automatically and consequently it takes about 30% lesser time during installing and dismantling compare with conventional form systems (Figure 10). After the utilization, the plastic forms can be cleaned with water and if it breaks, low voltage hot air gun is used to seal it. Other formwork materials such as wood, steel, aluminum produce various disadvantages which might outweigh their benefits. For example, the applications of wood are considerably costly and create substantial environmental impacts due to deforestation. With regard to aluminum formworks, it can be employed many times but its initial cost is high [7].

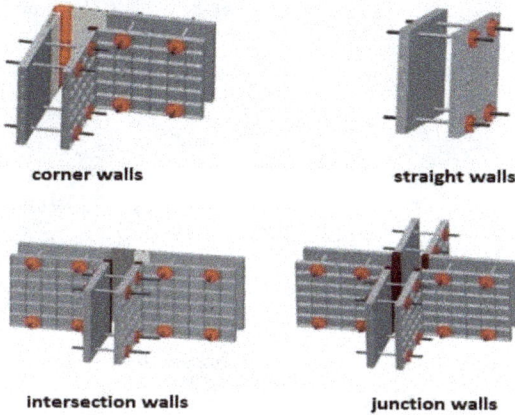

corner walls

straight walls

intersection walls

junction walls

Figure 10. Plastic formwork system for different walls

Table 1 shows the comparison between Plastic, Traditional, and Steel Formworks.

Table 1.Comparison between plastic, traditional, and steel formworks.

Items	Plastic formwork	Traditional formwork	Steel formwork
Recycled	40%	No	10%
Water resistant	Yes	No	No
Deformation condition	No	Yes	Yes
Stripping process	Easy	Moderate	Difficult
Size	Any size can supply	Restricted	Restricted
Corrosion resistant	Excellent	Bad	Bad
Availability time	More than 100 times	8 times	100 times

The plastic interlocking and modular systems are used to build widely variable, but relatively simple, concrete structures as shown in Figure 11. The panels are lightweight and very robust. They are especially suited for low-cost, mass housing schemes.

Figure 11. Reusable Plastic Formwork system for mass housing

7.7   Permanent Insulated Formwork

This formwork is assembled on-site, usually out of insulating concrete forms (ICF). The formwork stays in place after the concrete has cured, and may provide advantages in terms of speed, strength, superior thermal and acoustic insulation, space to run utilities within the EPS layer, and integrated furring strip for cladding finishes. Insulating Concrete Forms (ICF) are stackable, hollow, polystyrene blocks into which concrete is poured to form walls in buildings (Figure 12). ICF walls have many advantages, including ease of construction, high thermal resistance, and thermal mass, and airtight construction. These attributes lead to lower energy consumption in heating or cooling and increased comfort. Using polystyrene molds in place of timber or steel forms effectively kills two birds with one stone. It acts as form-work to mold the concrete and is then left in place to provide insulation for the walls (Figure 13). This building system is known generically as Insulated Concrete Formwork [8].

Figure 12. View of standard ICF Block (1010 x 250 x 270mm)

Figure 13. ICF wall panels, interlocked, and ready for concreting

7.8 Coffor

Coffor is a structural stay-in-place formwork system to build constructions in concrete. It is composed of two filtering grids reinforced by vertical stiffeners and linked by articulated connectors that can be folded for transport. A standard panel 1.10 m x 2.70 m (3' 8 x 9) weighs 32.7 kg (72 lbs) and can be carried by hand. After Coffor is placed, concrete is poured between the grids: excess water of concrete is eliminated by gravity, and the air is also eliminated. Coffor remains in the construction after the concrete is poured and acts as reinforcement (Figure 14). Any type of construction can be built with Coffor: individual houses, multi-story buildings including high-rise buildings, industrial, commercial or administrative buildings. Several types of civil works can be done with Coffor. Coffor is delivered completely assembled from the factory. No assembly is necessary on the construction site [9].

Figure 14. Coffor formwork system for the roof slab

7.9   Stay-In-Place structural formwork systems

This formwork is assembled on-site, usually out of prefabricated fiber-reinforced plastic forms. These are in the shape of hollow tubes and are usually used for columns and piers. The formwork stays in place after the concrete has cured and acts as axial and shear reinforcement, as well as serving to confine the concrete and prevent environmental effects, such as corrosion and freeze-thaw cycles [10].

7.10   Metal Beam Slab Formwork

Similar to the traditional method, but stringers and joists are replaced with aluminum forming systems, or steel beams and supports are replaced with metal props. This also makes this method more systematic and reusable. Aluminum beams are fabricated as telescoping units which allows them to span supports that are located at varying distances apart. Telescoping aluminum beams can be used and reused in the construction of structures of varying size.

7.11   Modular Slab Formwork

These systems consist of prefabricated timber, steel, or aluminum beams and formwork modules. Modules are often no larger than 3 to 6 feet or 1 to 2 meters in size. The beams and formwork are typically set by hand and pinned, clipped, or screwed together. The advantages of a modular system are: does not require a crane to place the formwork, speed of construction with unskilled labor, formwork modules can be removed after concrete sets leaving only beams in place before achieving strength.

7.12   Table or Flying Form Systems

Table Form shuttering is a kind of shuttering specializing in floor concreting and it is widely used in high building and skyscraper, multilayer industrial factory building, substructure, etc. These systems consist of slab formwork "tables" that are reused on multiple stories of a building without being dismantled. The assembled sections are either lifted per elevator or "flown" by crane from one story to the next. Once in position, the gaps between the tables or table and wall are filled with "fillers". They vary in shape and size as well as their building material. The use of these systems can greatly reduce the time and manual labor involved in setting and striking the formwork [3]. Their advantages are best utilized by large areas and simple structures. It is also common for architects and engineers to design the building around one of these systems (Figure 15).

Figure 15. A table or flying formwork system

Advantages of Table Form Shuttering

a) The table system is a Simple and Safe tool for Slab Construction that increases speed, time efficiency and cost economics on site.

b) Minimal System components in the table system make it a quick and easy to assemble system even by relatively unskilled labor.

c) Since no leg ties are required in the table system, areas below the soffit are kept clear ensuring easy movement of site operatives and machinery within the structure, which maximizes safety.

d) With a leg capacity of 35kN, larger tables can be constructed with various decking and propping options that enable the contractor to pour larger areas resulting in minimizing construction cycle time and cost.

e) The table formwork system offers good sustainability through reusability over many construction cycles with little waste.

f) The simplicity of assembly and disassembly results in large savings in time, effort, manpower and labor costs.

g) With the basic assembly of the table system being possible to be handled manually, savings are achieved by minimizing the use of crane time.

h) Table System can be transported and stored quickly and easily with a combination of crane lifting and hydraulic double-action jack trolley movement.

i) High Practicality Onsite with a simple design makes the table system a preferred system of Contractors for use in large scale developments.

7.13   Tunnel Forms

A tunnel formwork system is the latest innovations in the formwork industry. The use of repetitive cellular structures to construct both horizontal and vertical elements together is something that has got the potential to revolutionize the construction industry. Tunnel formwork is a mechanized system for cellular structures. It is based on two half shells which are placed together to form a room or cell (Figure 16).Tunnel forms are a large, room size form that allows walls and floors to be cast in a single pour. With multiple forms, the entire floor of a building can be done in a single pour. With tunnel forms, walls and slab are cast in a single day. The structure is divided into phases. Each phase consists of a section of the structure that will cast in one day. The phasing is determined by the program and the amount of floor area that can be poured in one day. The formwork is set up for the day's pour in the morning. The reinforcement and services are positioned and concrete is poured in the afternoon. Once reinforcement is placed, concrete for walls and slabs shall be poured in one single operation. The formwork is stripped the early morning and positioned for the subsequent phase. The formwork is manufactured in a fully automated plant.

Figure 16. A tunnel formwork system

Tunnel forms require sufficient space exterior to the building for the entire form to be slipped out and hoisted up to the next level. A section of the walls is left un-casted to remove the forms. Typically castings are done with a frequency of four days. Tunnel form can accommodate room widths from 2.4 to 6.6m. When rooms are wider (up to 11m), a mid-span table is incorporated between the tunnels. The main component of the system is the half tunnel. Manufactured entirely from steel, including the face of the form, the half tunnel provides the rigidity and smooth face necessary to produce a consistently high quality finish to the concrete [3].Tunnel forms are most suited for buildings that have the same or similar cells to allow re-use of the forms within the floor and from one floor to the next, in regions that have high labor prices.They enable construction of walls and floors together which make the process ideally suited for both high and low rise housing. It is easy to clean and reuse, the use of tunnel form systems also enables high quality surface finishes. Engineers are also assured of high dimensional accuracy of structures. The repetitive nature of the construction work is another plus point with the Tunnel formwork system (Figure 17).

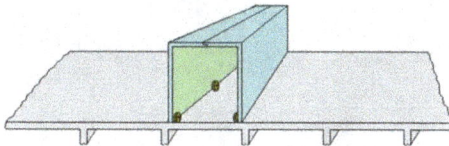

Figure 17. A repetitive tunnel formwork system for similar cells

7.14 Self-Climbing Formwork

Self-climbing formwork was conceived due to the huge construction projects that were required to be completed in tight schedules. Workplace safety is an important requirement that has also contributed to the development of self-climbing formwork techniques. These techniques have reduced the requirement of cranes. Workers now function in an enclosed safe environment, and there is adequate space for the tools and equipment. A self-climbing formwork presents an operational platform on which the concrete forms are placed, and the complete assembly is raised by a hydraulic cylinder (Figure 18). Each construction project has dissimilar characteristics, due to which the assemblies normally are customized to ensure their effectiveness and safety. Enhanced workplace safety, reduced construction period, and better efficiency are available when self-climbing formwork techniques are used [11].

Climbing formwork is usually used in the construction of buildings over five storey. Self-climbing, automated systems are generally used in the construction of buildings with more than 20-25 floors. Based on the site conditions, there are also instances when a combination of self-climbing and crane-handled jump form systems is used. The engineered nature of the formwork means that jump form systems allows for better control of the construction process. Repetitive use is possible adding to the cost-effectiveness of the construction process. Apart from offering enhanced safety, the use of jump form systems also ensures minimal concrete wastage and helps to stick to tight project deadlines [3].

Figure 18. A self-climbing formwork system

7.15   Flexible Formwork

In contrast to the rigid mold, flexible formwork is a system that uses lightweight, high strength sheets of fabric to take advantage of the fluidity of concrete and create highly optimized, architecturally interesting, building forms. Using flexible formwork it is possible to cast optimized structures that use significantly less concrete than an equivalent strength prismatic section, thereby offering the potential for significant embodied energy savings in new concrete structures. Concrete is a fluid that offers the opportunity to economically create structures of almost any geometry - we can pour concrete into a mold of almost any shape. This fluidity is seldom utilized, with concrete instead of being poured into a rigid mold to create high material use structures with large carbon footprints. The ubiquitous use of orthogonal mold as concrete formwork has resulted in a well-established vocabulary of prismatic forms for concrete structures, yet such rigid formwork systems must resist considerable pressures and consume significant amounts of material. Moreover, the resulting member requires more material and has a greater self-weight than one cast with a variable cross-section. Simple optimization methods may be used to design a variable cross-section member in which the flexural and shear capacity at any point along the element length reflects the requirements of the loading envelope applied to it [12].

By replacing conventional mold with a flexible system comprised primarily of low-cost fabric sheets, flexible formwork takes advantage of the fluidity of concrete to create highly optimized, architecturally interesting, and building forms. Significant material savings can be achieved. The optimized section provides the ultimate limit state capacity while reducing embodied carbon, thus improving the life cycle performance of the entire structure. Control of the flexibly formed beam cross-section is a key to achieving low-material use design. The basic assumption is that a sheet of flexible, permeable fabric is held in a system of falsework before reinforcement and concrete is added. Flexible formwork, therefore, has the potential to facilitate the change in design and construction philosophy that will be required for a move towards a less material intensive, more sustainable, construction industry [13]. The modern formwork systems listed above are mostly modular, which are designed for speed and efficiency. They are designed to provide increased accuracy and minimize waste in construction and most have enhanced health and safety features built-in.

Conclusions

Formwork is one of the most important factors in determining the success of a construction project in terms of speed, quality cost, and safety of work as it accounts for about 40% of the total project cost of the structure. The formwork sector is being influenced by the requirements of improved workplace safety and crane-independent systems that are fast and reliable. Working at heights has become less hazardous. Due to these reasons, concrete formwork is becoming popular. A formwork system signifies the molds used to store and hold wet concrete until curing is achieved. Curing is a significant process in concrete construction. Concrete has been demonstrated to be the most suitable construction material for buildings and foundations since it withstands fire and ensures protection against storms and extreme temperatures. Besides, concrete contributes to the creative modern architectural design. Therefore, modernization of formwork systems is essential to keep pace with concrete advancement. As tall structures have gradually emerged to be the modern trend, formwork systems have also developed from being simple timber-based to pre-engineered structures of steel, aluminum, timber, plywood, and plastics. The modern formwork technology has hence been translated into rapid construction with less labor.

References

[1] M. A. Shalgar, T. D. Aradhye, Introduction to advanced TUNNEL Formwork system: Case study of 'Rohan - Abhilasha', International Research Journal of Engineering and Technology (IRJET), 4(3), 2017.

[2] Information on http://www.concretecentre.com/technical_information/building_solutions/formwork.aspx

[3] RDSO, Report on Modern Formwork System, Report No. RDSO/WKS/2017/1, Research Design and Standards Organization, Lucknow, India, 2017.

[4] Information on https://www.constrofacilitator.com/plywood-formwork-advantages-types-and-applications/

[5] Information on https://theconstructor.org/building/formwork-shuttering/types-formwork-shuttering/3767/

[6] Information on https://happho.com/good-formwork-technical-functional-requirements/

[7]   Information on https://theconstructor.org/building/plastic-formworks-concrete-construction/15885/

[8]   ICF walls thermal performance, Research Highlight, Newsletter, New Delhi, 2007.

[9]   Information at https://civildigital.com/formwork-construction-types-applications-shuttering/

[10]   J. J. Orr, A. P. Darby, T. J. Ibell, M. C. Evernden and M. Otlet, Concrete structures using fabric formwork. The Structural Engineer, 89 (8), 2011.

[11]   Information on http://www.brighthubengineering.com/building-construction-design/49443-the-latest-formwork-systems/

[12]   K. Kostova,T. J. Ibell, A. P. Darbyand M. Evernden, Advanced composite reinforcement for fabric formed structural elements.2[nd]International Conference on Flexible Formwork, University of Bath, U. K. 2012.

[13]   J. J. Orr, A. P. Darby, T. J. Ibell and M. Evernden, Optimization and durability in fabric cast 'Double T' beams, 2[nd]International Conference on Flexible Formwork, University of Bath, U. K., 2012.

Novel Approaches in Materials and Technologies for Building Construction   Materials Research Forum LLC
Materials Research Foundations **158** (2024)   https://doi.org/10.21741/9781644902936

CHAPTER 10

Insulated Roof and Wall Tiles

1.   Introduction

The residential and commercial sectors account for 22% and 9% of the total electricity consumption respectively. With the rapid ongoing urbanization and economic development in the country, it has been estimated that India would build 700–900 million square meters of floor space per year for residential and commercial spaces in the next 20 years or so, leading to an extensive demand for electricity in the coming years [1]. Buildings can be planned so that residents enjoy minimally energy-intensive thermal and visual comforts. By starting with the early design stages of a building and using an integrated design approach, energy-efficiency measures can be efficiently incorporated. Such a strategy balances all energy use in a building; including lighting, air conditioning, and ventilation, while taking the local environment into account. In a hot country like India, the exterior building envelope surfaces, including the roof and walls, which are particularly exposed to direct solar radiation, heat up to temperatures greater than the normal inside temperature.

Thermal conduction transfers heat from the building exterior to the interior surfaces of the roof and walls, providing an unwelcome heat source inside the residents' area. Via convection and radiation, the heat is further disseminated inside the building, making the occupants very uncomfortable. As a result, it is crucial that the inhabitants' space possess the equipment needed to guarantee proper air circulation, ventilation, space cooling, and other necessities. Electrical fans and air conditioning systems typically accomplish these tasks. Along with heat loss due to leaks in the envelope from the warmer interiors to the cooler exterior, buildings in cold climatic regions also lose heat through conduction through walls and roofs. If this heat transmission is not well managed, a lot of heating will be needed to keep the inhabitants' environment at a comfortable temperature.

The skin of a building is called the building envelope, and it is supported by the structural framework of the building. Between the conditioned enclosed space and the outside environment, it serves as a thermal barrier through which thermal energy is transported. It is possible to significantly minimize the amount of energy required for space heating and cooling by limiting heat transfer through the building envelope [2]. Using an energy-efficient building design that uses thermal insulation in conjunction with other building materials in the building envelope will result in good thermal comfort at a low energy cost. Depending on their thermal conductivity rating, all building materials offer thermal resistance to the conduction of heat. However, compared to several building materials commonly used in construction, such as brick, having a thermal conductivity of 0.5 to 0.7 W/m°K, the thermal conductivity of insulation materials is remarkably low, often less than 0.06 W/m°K

Novel Approaches in Materials and Technologies for Building Construction   Materials Research Forum LLC
Materials Research Foundations **158** (2024)   https://doi.org/10.21741/9781644902936

2.   Role of Insulation on Thermal Performance of Buildings

Insulation of building envelopes, both opaque and transparent, is an important strategy for building energy conservation. The insulation strategy of a building needs to be based on a careful consideration of the mode of energy transfer and the direction and intensity in which it moves. The heat flow through a building component is proportional to its thermal conductance under steady-state conditions (U-value). Although outside circumstances change continuously to some degree, the steady state condition is strictly speaking never attained in practice, but the idea might be helpful in figuring out how well air-conditioned buildings function thermally in temperate and humid climates. In addition to its U-value, the thermal capacity of the element also determines the heat flow under non-steady state conditions, which are present in buildings without mechanical heating or cooling in all climates and even in air-conditioned buildings in hot-day climates. Under these conditions, 'thermal diffusivity', 'time-lag', and 'decrement factor' become significant [3].

Decrement factor and time lag are both characteristics of the construction element, not the building materials. Despite the fact that both may have the same U-value, the time lag is greater and the decrement factor is smaller for building elements of huge construction than for lightweight parts. The most significant areas for heat gain and loss are outside walls that are not insulated. Using the bulk of exterior walls will be more advantageous for insulation purposes [4]. It is possible to stop 70% of all heat loss by insulating the outer walls.

Insulating materials, due to their low thermal conductivity, can substantially resist the transfer of heat from the exterior to the interiors of the building if the external temperature is high (Figure 1) and resist heat transfer from the interiors to the exterior in a similar way when the external temperature is low. Figure 2 shows the effect of thermal insulation on heat transfer through walls. Figure 3 and Figure 4 show the details of Roof Insulation above and below the floor slab respectively. Figure 5 and Figure 6 show the wall section with external and internal insulation respectively.

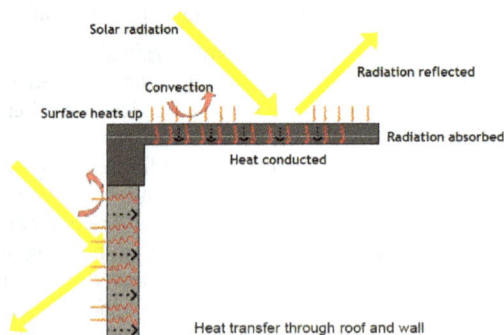

Figure 1. Effect of solar radiation on the building envelope and thermal conduction

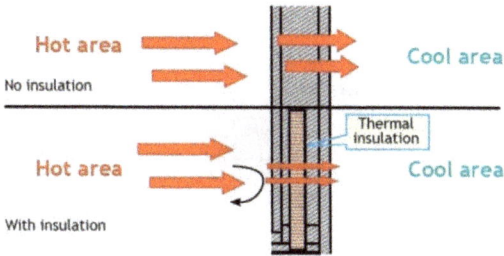

Figure 2. Effect of thermal insulation on heat transfer through walls.

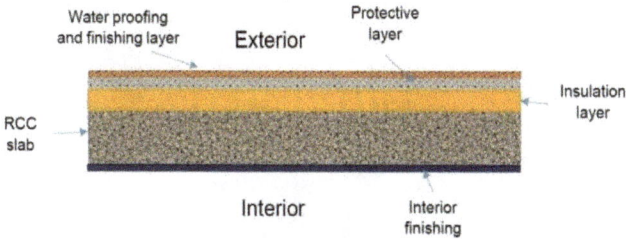

Figure 3. Details of roof insulation above the floor slab

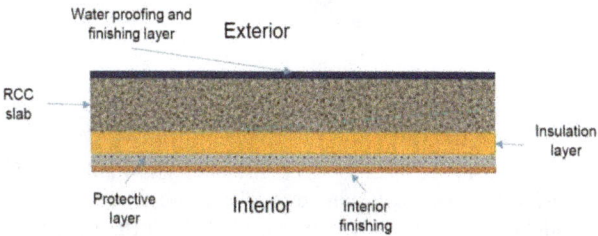

Figure 4. Details of roof insulation below the floor slab

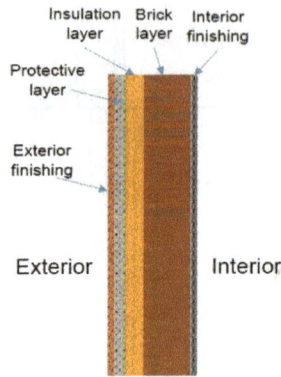

Figure 5. Wall with external insulation

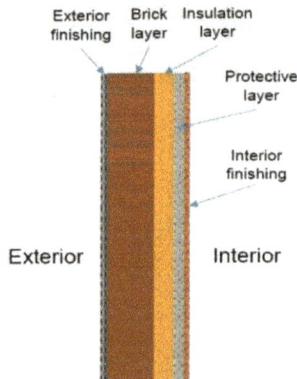

Figure 6. Wall with internal insulation

An insulating tile has been developed for deck insulation and walling. EPS is sometimes referred to as "molded expanded polystyrene" or "bead-board." This is moderate R-value insulation, from slightly less to slightly more than R-4 per inch, depending upon density [5].

3.   Expandable polystyrene (EPS) Kool tiles for walls and roof insulation

One such brand name is 'Kool tiles' developed in India by Reliable Insupacks P. Ltd. They are molded from a foamable plastic resin called Expandable Polystyrene (EPS). These tiles are lightweight as they are molded from polystyrene foam. Kool tile replaces the conventional mud-phuska laid over the concrete slab. The Kool tile is rectangular with a tongue and groove joint provided to enable locking together all tiles on their edges thus providing an airtight, monolithic insulation layer (Figure 7). The tile has an in-built surface, shaped for uniform bonding with screed or finishing layer. Top surface of the Kool tile is provided with lateral and longitudinal indentions of 10 mm depth to provide a shuttering surface for covering layer of mud or PCC for grouting the final/top layer of brick or other tiles [6].

Figure 7. Kool tile with the interlocking arrangement to achieve airtight insulation

Figure 8 shows the various dimensions for varying thicknesses of insulated EPS tiles of size 800 mm x 600 mm.

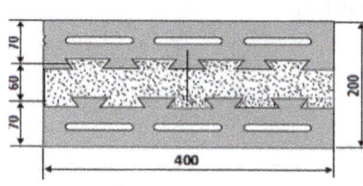

Figure 8. Sectional plan of a Kool tile

A difference in temperature upto 24°C can be achieved between the inside and outside temperature. Figure 9 shows the section of a tile wherein each layer of material assists in thermal resistance [7].

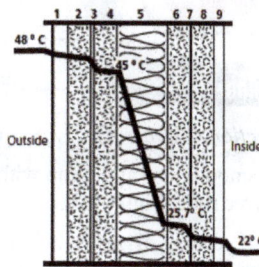

Figure 9. A Section of a Kool tile

4.   Manufacturing Process

a) This resin is foamed using saturated steam in a fully automatic Batch pre-expanding machine, to the specified density and conveyed pneumatically through a fluidized bed drier into air-permeable silos.

b) The beads are matured in these silos for a predefined maturation period and conveyed into material distribution bins for checking the density of these pre-expanded beads.

c) These beads after assessment of the density and manual intervention (by mixing higher or lower density beads (if required) are conveyed pneumatically into a hopper of a fully automatic molding machine.

d) An aluminum die with the required shape for the Kool tile is fitted onto this molding machine. Beads are drawn into this die as a pre-programmed sequence.

e) Steam passed through the beads to mold the beads into the required shape, the molding stabilized, die cooled and finally the Kool tile ejected onto a nylon slope for conveying to the finishing room (through a hot dry room).

f) After visual inspection of dry molded pieces, the product is packed in lots and shifted to the assigned store. Thereafter, inspection by the QC team is done and the material is dispatched.

5.   Installation Procedure

5.1   For External Roof

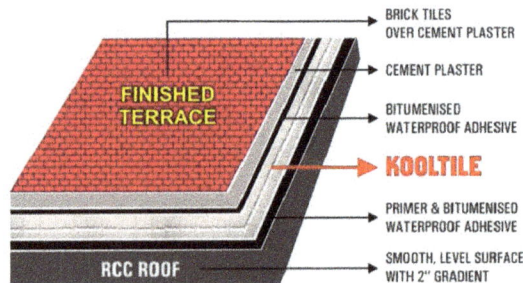

Figure 10. Section of a terrace covered with Kool tiles

a) Keep the roof surface smooth, dry and clean with a brush or with compressed air to make dust and dirt-free roof (Figure 10).

b) Apply uniformly two coats of asphalt-based polymerized waterproofing chemical or bituminous cold emulsion on the smooth surface (Figure 11).

c) Fix and join the tiles along with the slotted tongue and groove edges on the roof, press the tiles to make bitumen ooze out from the joist. Figure 3 shows the view of a roof kool tile with an interlocking arrangement to achieve airtight insulation by joining slotted tongue and groove edges.

d) Apply a base layer of cement screed over tiles for a minimum of 50 mm gradient. Fix brick tiles/mosaic / ceramic tiles over cement screed for terrace finish or any other finish of choice (Figure 12).

Figure 11. Kool sheets being laid on the terrace over a bituminous coat

Figure 12. Finished roof surface with Kool tiles

5.2   For External Wall

Figure 7 shows the different layers applied on the wall surface. The procedure for installing the Kool tiles on the wall is given as below:

a) Keep the wall smooth for protection and ease of adhesion. Apply uniformly two coats of bituminous cold emulsion on the smooth waterproof surface

b) Fix and join Kool tiles along with the slotted tongue and groove edges on the wall (Figure 13).

c) Hammer dowel-nails at all four ends to fix Kool tiles

d) Apply a base layer of cement screed over tiles for a minimum thickness of 4 mm. Apply paint / exterior finish on the plastered wall.

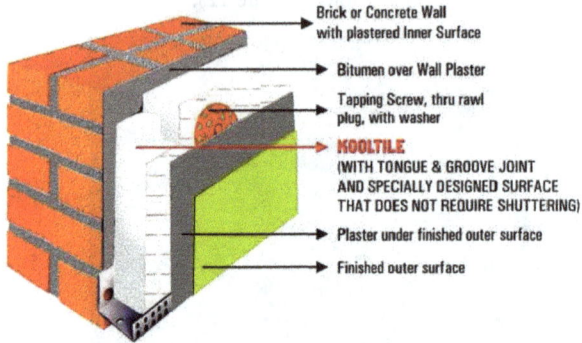

Figure 13. Section showing the procedure of installing Kool tiles on the wall surface

Figure 14. Kool tiles cladding on external walls

6.   Precautions while Laying the Tiles

Following precautions shall be taken while using these tiles on the roof:

a) The base of the roof slab/ wall shall be smooth for proper adhesion of the tiles, else air-pockets can lead to poor and uneven insulated surface

b) A minimum gradient of 1:120 shall be provided in the screed laid over the roof to avoid water logging which may lead to seepage and cracks on the roof.

c) Due to leaking water pipes/tanks etc. on the roof, rainwater outlets shall not block to avoid water logging which may lead to seepage and cracks on the roof. Proper grouting over brick tiles / ceramic tiles shall be laid. Tiles of lower density than specified by the manufacturer shall not be used as these may not be adequate to support heavy loads such as water tanks, movement of Diesel Generator sets, etc. on the terrace.

7.   Advantages of Insulated Roof and Wall Tiles

7.1   Roof Treatment

a) Individually molded roof insulation tiles give the entire roof a uniform insulation value.

b) Tiles are molded and not cut from an EPS block. Hence, each tile has got molded water repellent finish on both surfaces unlike the cut surfaces of sheets which may absorb some water through the porosities on the cut surface.

c) The benefits of a double layer are insulation provided by the interlocking joints of each tile to ensure water-tight joints of the concrete tiles, thus giving the perfect sealing effect.

d) There is a tremendous saving in costs. Other competing products with similar critical properties, like XPS or PUF, are three to five times as expensive.

e) The Comfort level of the top floor is significantly improved as the roof is insulated against the blistering heat in summers. There is 40-50% Electricity saving in the case an air conditioner is used. With 40% saving in a one-year payback of insulation, the cost is <1 year.

f) There are no aging cracks in the waterproofing of the roof (even for bitumen-based ones) and hence there is no seepage/dampness after heavy rains. The structure is protected due to the lifetime durability & reliability of roof insulation treatment with Kool-tiles.

7.2   External Walling

a) Provides an insulated weatherproof, durable architectural facade

b) It Helps conserves energy, reduces the building's overall CO_2 emissions

c) Help reduce heat loss or gain

d) Reduces the risk of condensation

e) Helps eliminate thermal bridging

f) Provides substantial reduction in sound transmission

g) Improves internal comfort levels even without AC's

h) Provides beneficial cost savings for life

i) Suitable for both new constructions and old

j) Reduces recurring investment in building maintenance

8.   Limitations of Use

a) Lower density tiles may not be adequate to support heavy loads such as multiple water tanks of capacity more than 5000 liters, movement of DG sets, etc. on the terrace.

b) Water logging may lead to seepage and cracks on the roof unless screed at the base is provided at a gradient of 1:120.

c) EPS, being a water vapor permeable material, may cause dampness and structural damage due to a cracked surface.

d) Limited options for using only cement concrete or high-end finishing materials as heat penetration stops at the insulation layer, leading to cracks on the finishing surface.

9.   Specifications of Kool Tile

9.1   Specifications of Kool Tiles for Walls

The varied criteria for selecting KOOL Tiles for wall insulation is listed as below [8]:

S. No.	Specifications	Kool Tile (Regular)	Kool Tile (HD)	Kool Tile (Supreme)
1	Thickness (mm)	37.5	37.5	37.5
2	Shuttering Thickness for Plaster	12	12	12
3	Constituent Material	EPS (Self Extinguishing)	EPS (Self Extinguishing)	EPS with Graphics (Self Extinguishing)
4	Color	White	White	Grey
5	Density (Kg/cum)	16	18	18
6	Thermal Conductivity (K Value) as per	0.37	0.35	0.31

7	IS:4671 (W/m^{20}C) at 10^{0}C (Each sqm of the roof exposed to say 200 C temp diff. (outside 450C) will cause a heat loss of 19.74 to 16.54 W)			
8	Thermal Conductivity (R Value) for Tile	0.987	0.933	0.827
9	Thermal Resistance (W/m^2 ^{0}C /W)	1.014	1.071	1.21
10	Water Absorption (% volume on 7 days immersion)	<0.6%	<0.1%	<1%

9.2   Specifications of Kool Tile for Roofs

The varied criteria for selecting KOOL Tiles for roof insulation are listed as below [8]:

S. No.	Specifications	Kool Tile (Regular)	Kool Tile (HD)	Kool Tile (Supreme)
1	Thickness (mm)	50	50	50
2	Density (Kg/cum)	16	32	40
3	Thermal Conductivity (K Value) as per IS:4671 (mW/m^{20}C) at 10^0 C	0.33	0.315	0.31
4	Thermal Conductivity (R-Value) for TIL (W/m^{20}C) at 10^{0}C (Each sqm of the roof exposed to say 200°C temp diff. (outside 45^{0}C inside 25^{0}C) will cause a heat loss of 12.4 to 13.2 W)	0.66	0.63	0.62
5	Thermal Resistance (m^{20} C /W)	1.515	1.587	1.613
6	Water Absorption (% Vol. on 7 days immersion)	<0.4%	<0.2%	<0.1%
7	Compressive Stress at 10 % strain (kPa)(wt. equivalent of 14-35 men (standing over 1 sqft area) reqd. to compress Kool Tile by 5 mm)	140	250	350

8	Sustained Compressive load bearing capacity with 2% (-1 mm compression) strain after 50 year (kPa)*	50	90	150

9.3 Dimensional stability

The optimal ability to retain volume and shape with changing temperatures.

9.4 Thermo stability

The short-term and long-term thermal stability of Kool Tiles are optimal.

9.5 Resistance to rotting/decaying and aging

Since it is not a naturally occurring organic product, there is no chance of decay or decomposition.

Conclusions

The operational energy consumption of buildings is greatly increased by heat transmission through roof slabs and exterior walls. Hence, in tropical regions, passive solutions are required to enhance the thermal performance of roof slabs. This essay has discussed the significance of using EPS insulated tile panels to increase the thermal performance of roof slabs and wall efficiency. It is crucial to lower the cooling load by employing insulating materials in the walls and roof in order to save money by using less electricity to cool the area. In order to make decisions about numerous architectural aspects, architects and building designers consider the thermal performance of a building envelope design. One of the largest and most important expenditures made by building owners is the construction of roofs. These significant expenses often include insulation, a waterproofing membrane, a roof deck, repairs, maintenance, and installation. The usage of expanded polystyrene, often known as EPS, in roofing systems has been recognized by roof system designers and contractors as having the ability to meet extremely strict building criteria. Closed-cell, robust, lightweight expanded polystyrene insulation is comprised of plastic foam. When built correctly and shielded from the effects of moisture, it can provide dimensional stability, strong water resistance, a long-term R-value, and significant savings on energy costs. Insulation made of expanded polystyrene pairs well with all commercial roofing systems. Modified bitumen, built-up roofs, fully adherent, mechanically fixed, or ballasted one-ply membrane systems are some examples of these expanded polystyrene-compatible roofing systems [9].

References

[1] Mckinsey, Global Institute. India's Urban Awakening: Building Inclusive Cities, Sustaining Economic Growth, 2010.

[2] M. Arif Kamal, F. Bano, Examining the Role of Building Envelope for Energy

Efficiency in Office Buildings in India, Architecture Research, 6(5), pp. 107-115, 2016.

[3] M. Arif Kamal, Material Characteristics and Building Physics for Energy Efficiency, Materials and Construction Technologies for Sustainable Development, Key Engineering Materials, 666, pp. 77-88, 2016.

[4] S. Bryant, E. Lume, The Bryant Walling System. Concrete '97 for the Future 18th Biennial Conference, Adelaide Convention Centre, pp. 641-649, 1997.

[5] T. Smith, Roofing systems, 2016, Information on https://www.wbdg.org/guides-specifications/building-envelope-design-guide/roofing-systems

[6] P. K. Adlakha, New Building Materials, and Technologies, Vol. IV, Indian Building Congress, New Delhi, India, 2019.

[7] Kooltile, Polystyrene Based Insulation Tile: Kooltile, Building Materials and Technology Promotion Council, New Delhi, India, 2014, Information on https://bmtpc.org/DataFiles/CMS/file/PDF_Files/30_PACS_kooltile.pdf

[8] Reliable Insupacks (P) Ltd., Greater Noida, A visual guide for over deck roof insulation, 2018.

[9] Polymolding, 2022, The Expanding Role of Expanded Polystyrene in Roofing Construction, Information on https://www.polymoldingllc.com/the-expanding-role-of-eps-expanded-polystyrene-in-roofing-construction/

Novel Approaches in Materials and Technologies for Building Construction   Materials Research Forum LLC
Materials Research Foundations **158** (2024)   https://doi.org/10.21741/9781644902936

CHAPTER 11

Metal Roofing System for Warehouses

1.   Introduction

A warehouse is a storage building usually characterized as a single storey steel structure with or without mezzanine floors. The designing of warehouse includes designing of the structural elements including principal rafter and roof truss, column and column base, purlins, sag rods, tie rods, gantry girder and bracings. The roofing innovations in warehouse and industrial buildings address the issues raised by traditional materials, be it installation time, labor effort, durability, sound and temperature insulation, design flexibility and many more aspects. Metal roofs in a warehouse and industrial buildings were made first from copper or tin in early times. In the 19th century there were roofing materials like galvanized iron through aluminum, but now steel has taken a revolutionary step for fast completion of buildings and offering recycling benefits. Structurally metal roofing for a warehouse can be classified as structural and non structural or architectural [1]. Whereas structural metal roofing is directly attached to purlins and eliminates the need of any solid support beneath, non structural metal roofing requires some solid support beneath in the form of plywood, a metal roof deck etc. The non-structural systems would include tiles, corrugated metal roofing as well as panels. Metal roofs are available in sheets and sections (Figure 1). Sheets which come in standard sizes can be cut whereas sections are usually custom built. Custom made panels made of high quality metal roofing are fastened with screws and rubber washers and can simulate looks of slate, tiles etc. (Figure 2).

Figure 1. View of metal roofing sheets

Figure 2. Exposed fastener system in metal roofing of warehouses

Standing Seam Metal Roofing (a commonly used sheet metal roof), besides imparting an appealing finished look to the roof, conceals the washers and screws by using clips. The brand names like Tata BlueScope, Kingspan, Colour Roofs (India), Everest Industries and Multicolor Steel emerged as leaders in roofing sheets in steel, galvalume (55% Aluminum-Zinc alloy coated sheet steel) and other materials. All these have the advantage of high strength profiling, corrosion resistance with multiple coatings and insulation panels adding to interior comfort (Figure 3). On site roll forming technology and fixing large length sheets for industrial sheds were pioneered in the country. The capability of spanning larger areas per unit mass is another reason for the product acceptance. Interarch's Tracdek for example, scores with its Galvalume steel products with organic coatings that face the challenges of corrosion even in coastal areas [2].

Figure 3.View of metal roofing in warehouses

2.   Types of Materials for Metal Roofing System for Warehouses

There are a multitude of different metal materials suitable for warehouse roofing construction including steel, aluminium, and zinc. These materials are available as both metal roof sheets and metal roof tiles. Metal is commonly utilized throughout the industry for its impressive longevity, top durability, and industrial look. The most commonly used metal roofing materials for warehouses include steel, aluminum, copper, zinc, and titanium. Metal alloys in different colours, styles and textures are also available. The different types of metals that can be used for warehouse roofing are summarized as below:

a) **Copper:** The most expensive of all materials, the roofing is beautiful and is generally used in smaller places. Though prone to weathering to a green colour with time, it is resistant to corrosion and does not need to be painted. Care has to be taken to avoid direct contact with other metals. It has value in recycling.

b) **Aluminum:** Aluminium is a very preferred choice for the roofing of warehouses. is lightweight, durable, ductile, malleable and corrosion resistant. It has a light reflectivity of over 80% and is used to a great deal in industrial buildings. One needs to avoid direct contact with dissimilar metals, concrete or mortar; and moisture entrapment during storage to avoid staining.

c) **Zinc:** Resistant to corrosion, it's durable (a lot more than steel too) and can be recycled. The only downside is the very expensive initial cost.

d) **Steel:** One of the most common, steel is an inexpensive material. Its strength and its prevention from rusting by using a zinc coating, a sealant or some paint are some of its advantages. Steel sheets for roofs can be manufactured from colour coated Galvalume / Galvanized, Soft Steel and High Tensile Steel.

e) **Titanium:** It has high strength, is light weight and immune to atmospheric corrosion.

f) **Tin:** Commonly seen in Indian residential metal roofing they are a great prevention from the winds.

g) **Stainless Steel:** With a lustrous finish it doesn't need any extra coatings, but its built-in anti corroding properties make it expensive. After installation, the surface needs to be cleaned to remove contaminations that can lead to corrosion. Space frames are popular in stainless steel.

h) **Galvanized Steel:** It is steel coated with zinc which is often used in commercial and industrial applications. It is economical, can come in direct contact with concrete and masonry, is light weight and durable. It needs to be insulated with bituminous coating if it's going to come in direct contact with copper.

i) **Alloys:** Different metal alloys are good for metal roofing; the cost however depends on what the alloys consists of. The Terne metal (zinc-tin alloy to cover steel) is light weight, durable and has low expansion though it cannot be nailed through metals and must be painted soon after installation (Figure 4). Also, it needs to be primed on both sides before installation. Then there are alloys like galvalume or zincalume (blend of zinc, aluminum and silicon-coated steel) which are more widely used as a base metal under factory coated colours.

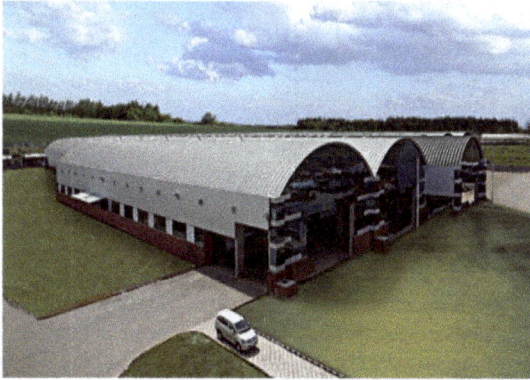

Figure 4. Self-supporting steel roofs impart a clean and dignified look to the warehouse

3. Various Metal Roofing Systems for Warehouses

There are different types of roofing systems that may be used in warehouse and industrial buildings. These tend to fall into some broad categories, which are described in the following sections [3].

3.1 Single-skin trapezoidal sheeting

Single-skin sheeting is widely used in agricultural and industrial structures where no insulation is required. It can be used on roof slopes down to as low as 4° provided that the laps and sealants are as recommended by the manufacturers for shallow slopes. The sheeting is fixed directly to the purlins and side rails, and provides positive restraint (Figure 5). In some cases, insulation is suspended directly beneath the sheeting.

Generally steel sheeting is made of galvanized steel grades S280G, S320G or S275G to EN 10326. Due to the wide range of product forms, no standard dimensions for sheeting exist. The sheets are usually between 0.50 and 1.50 mm thick (including galvanization).

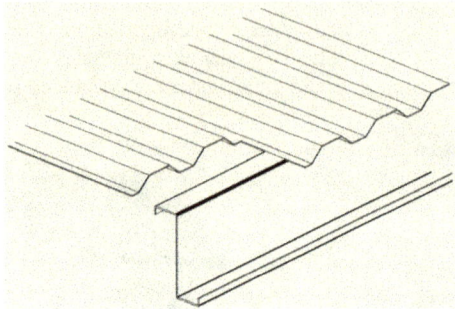

Figure 5.Single-skin trapezoidal sheeting

3.2   Double Skin System

Double skin or built-up roof systems usually use a steel liner tray that is fastened to the purlins, followed by a spacing system (plastic ferrule and spacer or rail and bracket spacer), insulation and an outer sheet. Because the connection between the outer and inner sheets may not be sufficiently stiff, the liner tray and fixings must be chosen so that they provide the level of restraint to the purlins. Alternative forms of construction using plastic ferrule and Zor rail and bracket spacers are shown in Figure 6 and Figure 7. As insulation depths have increased to provide greater insulation performance, there are 'rail and bracket' solutions, as they provide greater stability. With adequate sealing of joints, the liner trays may be used to form an airtight boundary.

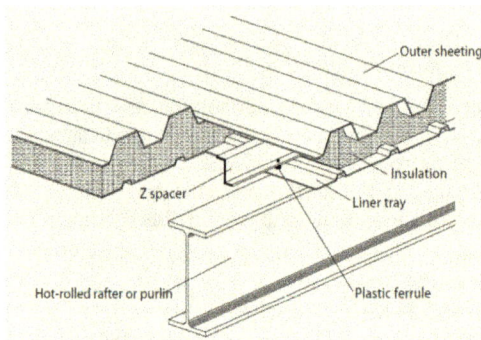

Figure 6. Double-skin construction using trapezoidal ferrule and Z spacers

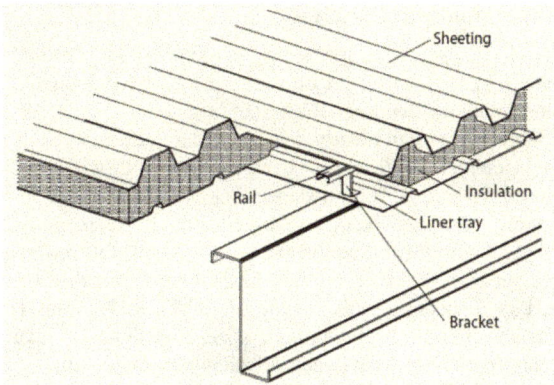

Figure 7. Double-skin construction using rail and bracket spacers

3.3 Standing Seam Sheeting

Standing seam sheeting has concealed fixings and can be fixed in lengths of up to 30 m. The advantages are that there are no penetrations directly through the sheeting that could lead to water leakage, and fixing is rapid. The fastenings are in the form of clips that hold the sheeting down but allow it to move longitudinally (Figure 8). The disadvantage is that significantly less restraint is provided to the purlins than with a conventionally fixed system. Nevertheless, a correctly fixed liner tray will provide adequate restraint to the purlins.

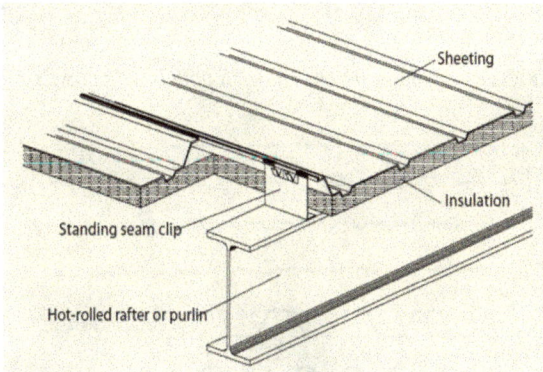

Figure 8. Standing seam panels with liner tray

3.4 Composite or Sandwich Panels

Composite or sandwich panels are formed by creating a foam insulation layer between the outer and inner layer of sheeting. Composite panels have good spanning capabilities due to composite action in bending. Both standing seam (Figure 9) and direct fixing systems are available. These will clearly provide widely different levels of restraint to the purlins. Sandwich elements for roofs generally have a width of 1000 mm with thicknesses between 70 and 110 mm, depending on the required insulation level and structural demands. Despite being relatively thick elements, self-weights are comparatively low. Component lengths of up to 20 m for roofs and walls permit constructions without or with few joints. The basic material for the outer layers is usually galvanized coated steel sheeting with thicknesses of 0.4 to 1.0 mm. The inner layers of sandwich panels are often lined or slotted; special designs are available with plane surfaces. Some patterns of sandwich panels are shown in Figure 10.

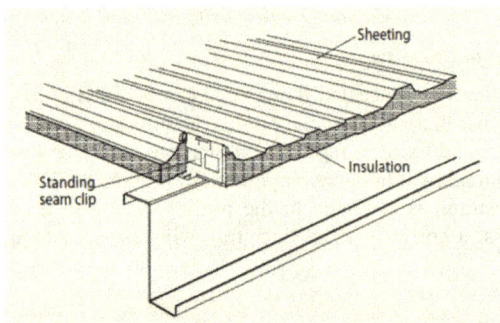

Figure 9. Composite or sandwich panels with clip fixing

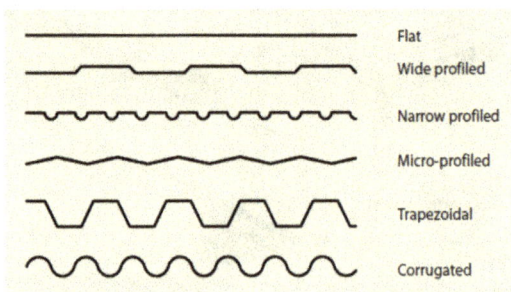

Figure 10. Types of external surfaces for sandwich panels

4.   Advantages and Disadvantages of Metal Roofing for Warehouse

4.1   Advantages of Metal Roofing in a Warehouse

The advantages for using metal roofing in a warehouse are:

a) Low end metal roofing lasts 20 years while some sheet materials can last up to 100 years. Most steel roofing comes with a 50-year guarantee, when installed correctly. Not only can metal roofing last decades, but more importantly, its low-maintenance. During this time it provides robust and hard-wearing protection against high winds, heavy storms, deep snows, mildew, insects, and rot. There are high quality paint systems/coatings used on metal roofing to resist chalking and fading of their colors. Simultaneously, these paint systems help in shedding off the dirt and resist the growth of fungal matter leading to low maintenance too.

b) Despite its impressive durability, metal roofing is actually pretty lightweight which makes for far easier transportation, handling, and installation. This reduced weight also helps you minimize costs for engineering and the construction of the supporting structure.

c) In severe wind conditions, it has been found that metal roofs outlast other roofing products. Also, metal roofing is ideal in northern or mountainous regions since painted metals roofs help in snow shedding.

d) With coatings of various colors and patterns, and availability in a number of profiles, metal roofing offer extensive design flexibility to architects.

e) There is no denying the strength of metal roofing and they are amongst the lowest weight roofing products - a benefit in seismic areas and in retrofitting projects. This also reduces the load on the structure.

f) Most metals can be recycled without losing any valuable properties leading to eco-friendliness. Most metal roofing materials are constructed using anywhere from **25%** up to **95%** recycled content, making them highly sustainable and environmentally-friendly. Metal roofing also doesn't contribute to landfill waste as it is 100% recyclable itself at the end of its lifespan. Aluminium and copper roofing products are generally available with 95% recyclable content.

g) Installation and construction is easy and quick and in many cases metal roofing can be installed over existing roofs also, saving costly tear off. Its light weight and ease of installation means that it can be safely installed in situations where other materials wouldn't be feasible.

h) Metal roofs reflect heat effectively, thus minimize the heat of absorption leading to lowering of air conditioning costs.

i) Metal roofing has great fire resistance. Many manufacturing businesses work with a much higher fire risk than other businesses. As such, they need to take extra precautions to stay fire-safe. Metal roofing is 100% non-combustible, making it the perfect choice for industrial buildings. Metal roofing accommodates any heat-related contraction and expansion, minimizing the risk of collapse should a fire break out.

j) Metal roofing in warehouses saves interior spaces. The space is crucial for any business, but it's especially important for warehouses. A higher ceiling means taller shelving and more space to fit stock. Unlike other roofing materials, metal is lightweight. As a result, it can be installed with minimal scaffolding and supports, freeing up additional warehouse space. This also allows air conditioning systems to be installed higher up rather than hanging down and interfering with your shelving [4].

4.2   Disadvantages of Metal Roofing in a Warehouse

The disadvantages of using metal roofing in a warehouse are summarized as below:

a) The difference in temperature or the sound of falling rain sometimes can produce noise like in case of curved or corrugated metal roofing. This can however be avoided by proper installation and using roofing material that contains structural barriers to minimize the sound effect.

b) Initial cost is expensive. Their costing is on average around two to three times more than asphalt shingles, metal roofing is pretty pricey. When used in large scale industrial warehouse projects, metal roof materials will add up to a considerable total cost but may be considered worthwhile in the long-run because of their durability.

c) Metal roofing are noisier than others types of roofing. The metal roofs can get pretty loud when exposed to heavy rain or hailstorms. To minimize sound transmission, additional layers of sheathing or insulation can be installed, although this will entail a further cost to the project.

d) Dissimilar and incompatible materials can cause unexpected and rapid corrosion.

e) Transport of metal roof components is an energy intensive activity.

f) Metal roofing renders the interiors warmer but Fabian mentions that with the use of an insulating layer (air-ventilation, insulation material such as Styrodur, Thermocol etc) one can easily control the radiation to the interior.

g) When installed in large sheets, metal roofing can be quite challenging to repair as full panels will need to be removed and replaced. Finding replacement sheets that perfectly match the shade of the existing roofing can also be pretty difficult.

Conclusions

Metal roofing is used in long span, column free structures such as stadiums, airport hangers, etc and especially in industrial buildings and warehouses. The applicability of metal or steel roofing in warehouses has increased since the insulation and ventilation solutions have also paired up. Low cost, light weight and low maintenance were and still are the popularity pushers in a warehouse. This exceptional durability is what makes a metal roof so practical for industrial warehouses in need of a reliable, long-term roofing solution. Modern technology has added precision, safety, insulation, lifetime looks and textures to up the style quotient. There is also considerable reduction in both labor effort and material used. Steel roofing companies also train workers on installation, further assuring delivery of quality of the installation. The after sales service and wide spread dealer networks also contributed in a large way for the growing popularity of this roofing. With standard factory produced sheets and accessories, repair and replacement now no longer means obvious patchwork on the exterior. Metal roofs also add value to the warehouse in being fire retardant and scoring for green architecture.

References

[1]   Information on https://www.americanweatherstar.com/the-5-most-common-types-of-warehouse-roofing/

[2]   Information on http://my.whirlwindsteel.com/blog/bid/407623/Steel-Warehouse-Projects-5-Key-Benefits-of-Building-with-Metal

[3]   Information on http://nbmcw.com/articles/peb-roofing/roofing-cladding/21471-the-new-age-of-metal-roofing.html

[4]   Information on https://www.watermasterroofing.com.au/why-metal-roofing-is-the-best-choice-for-warehouses-and-factories/

About the Authors

Dr. Mohammad Arif Kamal is an architect and academician having around 20 years of Teaching, Research and Professional experience in the field of Architecture and Building Construction Technology. He was awarded a Ministry of Human Resource Development (MHRD) fellowship for pursuing both Master of Architecture (M. Arch.) and Ph.D. degree in Architecture from Indian Institute of Technology (IIT), Roorkee, India. Dr. Kamal is presently working as an Associate Professor in Aligarh Muslim University, India. His area of research includes Environmental Design, Climate Responsive Architecture, Sustainable Architecture, and Building Construction Technology etc. He has published around 100 research papers in various international journals and conferences. He has published 9 books and 15 book chapters. Dr. Kamal is Editor-in-Chief of 7 International journals related to Architecture, and Building Science and Technology. He has also edited 5 Special Topic Volume (Scopus indexed) related to Sustainable Building Materials, published by Trans Tech Publications, Switzerland.

Rupesh Manohar Surwade received his Master of Design (M. Des.) in Industrial Design from the School of Planning and Architecture (SPA), New Delhi, India. Presently Rupesh is an Associate Professor at Priyadarshini Institute of Architecture and Design Studies, Nagpur, India. He is an architect and urban planner having 17 years of teaching, research and professional experience in the field of designing, planning, building construction techniques, rendering techniques, and various manufacturing methods. His research interests include Architecture, Retail Design, Furniture Design, Product Design & development.

Abhishek Mahendra Bangre did his Masters in Industrial Design (M. Des.) from RTM Nagpur University, India. Presently Abhishek is working as an Assistant Professor at Priyadarshini Institute of Architecture and Design Studies, Nagpur India. He is an active researcher in the fields of product design and architecture with more than 15 years of experience as a practicing architect.

www.ingramcontent.com/pod-product-compliance
Lightning Source LLC
Chambersburg PA
CBHW071655210326
41597CB00017B/2213